Food Science and Technology

nova
science publishers
New York

Food Science and Technology

Food Processing: Advances in Research and Applications
Myriam Huijs (Editor)
2022. ISBN: 978-1-68507-570-5 (Hardcover)

A Closer Look at Polyphenolics
Peter Bertollini (Editor)
2022. ISBN: 978-1-68507-434-0 (Hardcover)
2022. ISBN: 978-1-68507-480-7 (eBook)

Food Processing: Advances in Research and Applications
Myriam Huijs (Editor)
2022. ISBN: 978-1-68507-570-5 (Hardcover)
2022. ISBN: 978-1-68507-581-1 (eBook)

Cheeses around the World: Types, Production, Properties and Cultural and Nutritional Relevance
Raquel de Pinho Ferreira Guiné, PhD, Paula Maria dos Reis Correia, PhD and Ana Cristina Ferrão, Msc (Editors)
2019. ISBN: 978-1-53615-418-4 (Hardcover)
2021. ISBN: 978-1-53615-419-1 (eBook)

Nutraceuticals: Food Applications and Health Benefits
Anita Kumari, PhD (Editor) – Assistant Professor, Department of Nutrition Biology, Central University of Haryana, Mahendergarh, Haryana, India
Gulab Singh, PhD (Editor) – Assistant Professor, Maharaja Agrasen University, Baddi, Himachal Pradesh, India
2022. ISBN: 978-1-68507-434-0 (Hardcover)
2021. ISBN: 978-1-68507-512-5 (eBook)

More information about this series can be found at https://novapublishers.com/product-category/series/food-science-and-technology-series/

Myriam Huijs
Editor

Food Processing

Advances in Research and Applications

nova
science publishers
New York

DOI: https://doi.org/10.52305/BEEN9615

Library of Congress Cataloging-in-Publication Data

ISBN: 978-1-68507-570-5

Published by Nova Science Publishers, Inc. † New York

Contents

Preface

This volume includes seven chapters, each of which detail recent advances in research and applications of food processing. Chapter one describes symbiotic foods and their effects on consumer health, mainly on obesity and gastrointestinal diseases, covering their mechanisms of action, types, application, and legislation, as well as the impact that probiotics and prebiotics have on diseases. Chapter two analyzes current data on different polyphenol nanoformulations as alternative chemotherapeutic and chemopreventive treatments of cancer. Chapter three identifies the phytochemicals present in purslane, its health benefits, and its applications in the food industry. Chapter four explores the potential of *Amaranthus* spp. as a high nutritional value food alternative. Chapter five describes the fundamentals of refrigeration, its types, recent developments and multi-dimensional applications in food industries. Chapter six synthesizes some fundamental aspects of the relationship between bee pollen's constituents and its bioactivity, in light of recent publications which reveal its potential as a healthy product. Finally, chapter seven explores the potential of aguamiel as a healthy option in the food industry.

Chapter 1 - Currently, diseases related to wrong eating habits, lack of exercise, and accelerated lifestyle have increased worldwide. Therefore, adjuvant therapies are currently being developed to treat some pathologies that include the consumption of symbiotic foods, that contain probiotics and prebiotics. Because, in recent years, the number of gastrointestinal diseases, such as irritable bowel syndrome, allergies, food intolerances, and inflammatory bowel disease has increased, since these diseases have in common an affected microbiome. Thus, eating symbiotic foods could improve patient prognosis. Accordingly, this document describes the relevant information on symbiotic foods and their effects on consumer health, mainly on obesity and gastrointestinal diseases, covering their mechanisms of action, types, application, and legislation, as well as, the impact that probiotics and prebiotics have on diseases, finally, a reflection is made of the future trend of symbiotic foods in the food industry.

Chapter 2 - Cancer is defined as a group of multifactorial diseases caused by environmental and genetic factors that induce a series of mutations in normal cells that cause abnormal cell growth with possible malignant and fatal neoplasties. Polyphenols are natural phytochemicals that have been widely used for the prevention and treatment of various ailments due to their anti-glycemic, anti-inflammatory, antioxidant, anticancer, and cytotoxic effects; thus, they could potentially be capable of acting as chemotherapeutic and chemopreventive agents in the treatment of cancer. However, these phytochemicals have low bioavailability. Therefore, nanoparticles (NPs) are likely to provide an attractive new alternative to conjugate the efficacy of various bioactive phytochemicals in cancer therapy. This is due to the physicochemical, optical, magnetic, electrical, mechanical properties, surface area, size, and dispersion that these nanostructures have, which would provide an innovative green nanotechnology approach to the use of polyphenols to obtain nanoparticles. In this review, the authors have analyzed the most current data on the different polyphenol nanoformulations as alternative chemotherapeutic and chemopreventive treatments in the treatment of cancer and discuss their multiple actions and anti-cancer molecular mechanisms that induce apoptotic potential in cancer cells.

Chapter 3 - Functional foods contain nutritional and biologically active compounds which provide health benefits, this type of food can be obtained naturally from different plants. One of these plants is purslane (*Portulaca oleracea* L.) which is distributed worldwide, mainly in England, Saudi Arabia, India, China, and some African countries, it is even well distributed throughout Mexico. In the last country, it is identified as an ancestral and traditional food, although in recent years, it is less consumed, mainly with the development of highly processed foods and those of fast preparation, so the nutritional properties and health benefits that this plant can provide to the human body are little known. On the other hand, obesity and cardiovascular diseases are some of the main health problems in different countries. Thus, it is essential to determine the impact of different foods on health, in order to develop an ingredient or functional food that helps to solve or prevent some of the health problems mentioned above. The objective of this document was the search and discuss recent literature on *Portulaca oleracea*. It was possible to identify the phytochemicals present in purslane, its health benefits, and its applications in the food industry.

Chapter 4 - Currently, amaranths are considered alternative crops with food potential and comprise more than 350 species. They have been used in human food for more than 1000 years. However, the culinary traditions that included them daily in the diet have been lost. This genus is distributed throughout the world and is characterized by its resistance to adverse climates, and in most countries, these plants are considered invasive. However, research in recent decades has demonstrated the importance of *Amaranthus* spp. as an alternative source of nutrients (protein, fiber, minerals, and carbohydrates), which could give a favorable turn in the diversity of agroecosystems, health, cultural richness, and technological innovation in the food industry. On the other hand, biodiversity encompasses different levels of organization that include composition, structure and function to understand the importance of highly valued edible plants, such as *Amaranthus* species. From *Amaranthus* plants, stems, leaves, flowers, and seeds can be used to obtain micro-macro nutrients and phytochemicals. In addition, extracts from their roots are suitable as bioherbicides. Additionally, it is recommended to study the future impact of technological and agricultural developments in food security in developing countries.

Chapter 5 - Refrigeration also known as artificial cooling is the process of exchanging heat from a source (an object) at high temperature to a sink at lower temperature obeying second law of thermodynamics. This chapter describes the fundamentals of refrigeration, its types, recent developments and multi-dimensional applications in food industries. The vapor compression cycle is discussed in details describing all its four components (evaporator, compressor, condenser & expansion valve) and how refrigerants flow between these components changing its state to produce refrigeration effect. This cycle is also explained with the help of pressure-enthalpy and temperature-entropy diagrams. This cycle gives an insight into how cooling is achieved at its maximum level while minimizing the amount of work done. Further, this chapter presents the different types of refrigerants and the way of designating those refrigerants. Also the operating concept of vapor absorption refrigeration circle and multistage refrigeration system is succinct emphasizing its distinctions cascade refrigeration system. This chapter also touches various household and industrial applications of these systems for the purpose of storage and post harvest management of food materials with technical elaborations.

Chapter 6 - Since ancient times, bee pollen has emerged as a rich food product, which, besides being a source of energy, contains beneficial substances with recognized properties, both nutritionally and therapeutically.

Recently, research about this natural product has allowed a more supported recognition of its health enhancement potential. The therapeutic properties of bee pollen are associated with the chemical composition, in particular vitamins, minerals, polysaccharides, and bioactive compounds, like phenolic acids and flavonoids. Due to the reasons highlighted, this review aims to synthesize some fundamental aspects of the relationship between the product's constituents and their respective bioactivity, in light of the most recent publications, which reveal the potential of bee pollen as a healthy product. Among them, citations of the most relevant benefits report the antioxidant, anti-inflammatory, antifungal, and antimicrobial bioactivities. Other potentialities, as the therapeutic use in drug discovery, include, for instance, the ability to modulate cancer development. Along the sections of this chapter these health effects will be presented and discussed.

Chapter 7 - Nowadays, the development of foods that promote health and nutrition is one of the priorities in the food industry. Although there is a great variety of products and beverages with a pleasant flavor and nutritional value that have been undervalued, among them is aguamiel (sap from the maguey *Agave* spp), which is a beverage that provides different nutrients and has a cultural tradition in some communities in Mexico. Aguamiel can be considered a unique and special beverage, especially because of its artisanal production process. However, this traditional beverage has been losing its cultural value. On the other hand, aguamiel is the sap of the maguey (*Agave* spp), which is considered a natural resource of great diversity of the Agavaceae family. The human population currently consumes nutritional products and beverages, some are of good quality and provide nutrients, but others do not, since they are highly processed foods and this causes the quality and nutrition of the product to be low. This reflects the importance of preserving the customs of a traditionally 100% Mexican beverage that is also healthy. In northern Mexico, although there is a tradition of consuming this drink, it could be said that its consumption is minimal, although studies have shown that this drink is healthy, with a delicious flavor, contains vitamin A, B, C and iron, proteins, carbohydrates and is an excellent antioxidant. In addition, this drink has therapeutic medicinal properties, and helps to lower cholesterol. That is why aguamiel is one of the products that has not been fully exploited, since the main use that has been given to it is for the production of pulque, a traditionally Mexican alcoholic beverage that results from the fermentation of aguamiel.

Chapter 1

Symbiotic Foods and Their Relationship to Gastrointestinal Diseases and Obesity

Stephany Nefertari Chávez García[1], Sonia Yesenia Silva Belmares[1], Sendar Nery Flores[1], Adriana Carolina Flores Gallegos[1], Xochitl Ruelas Chacón[2] and Raúl Rodríguez Herrera[1,*]

[1]School of Chemistry, Universidad Autónoma de Coahuila, Saltillo, Coahuila, México

[2]Food Science and Technology Department, Universidad Autónoma Agraria Antonio Narro, Saltillo, Coahuila, México

Abstract

Currently, diseases related to wrong eating habits, lack of exercise, and accelerated lifestyle have increased worldwide. Therefore, adjuvant therapies are currently being developed to treat some pathologies that include the consumption of symbiotic foods, that contain probiotics and prebiotics. Because, in recent years, the number of gastrointestinal diseases, such as irritable bowel syndrome, allergies, food intolerances, and inflammatory bowel disease has increased, since these diseases have in common an affected microbiome. Thus, eating symbiotic foods could improve patient prognosis. Accordingly, this document describes the relevant information on symbiotic foods and their effects on consumer

* Corresponding Author's E-mail: raul.rodriguez@uadec.edu.mx.

In: Food Processing
Editors: Myriam Huijs
ISBN: 978-1-68507-570-5

health, mainly on obesity and gastrointestinal diseases, covering their mechanisms of action, types, application, and legislation, as well as, the impact that probiotics and prebiotics have on diseases, finally, a reflection is made of the future trend of symbiotic foods in the food industry.

Keywords: probiotics, prebiotics, intestinal inflammation, irritable bowel, allergies, food intolerances

Introduction

Wrong eating habits have been reflected in health problems, such as obesity and gastrointestinal diseases. The prevalence of these diseases has increased in recent decades in different countries. Obesity is a pathology where excessive fat accumulates, affecting metabolically the human body (Calderón and Acosta, 2019). While gastrointestinal diseases are characterized by affecting the function of the organs of the digestive system and the body's metabolism, they are generally caused by bacteria, parasites, viruses, and the consumption of some foods. Among these diseases are inflammatory bowel disease, irritable bowel syndrome, and bacterial overgrowth in the small intestine (Nesterova et al., 2020). Interventions in this type of disease include medications and a specific diet, with nutrients, food, and supplements. In Mexico as in other developing countries, obesity is considered an epidemic that includes 36.1% of the population, which increases the risk of comorbidities; thus, the problem must be addressed, immediately (ENSANUT, 2018). Besides, the increase in these diseases affects human health and consumes a large number of resources destined for public health (Campos et al., 2018). These pathologies have in common, a condition in a greater or lesser proportion of the intestinal microbiota, which has led to research on the benefits of consuming probiotics, prebiotics, and symbiotics, considered as functional components, which can enhance specific and beneficial functions in the human body (Acker and Cash, 2017).

In this context, probiotics can be defined as living microorganisms which in adequate amounts confer a benefit on the consumer's body (FAO/WHO, 2001). While, a prebiotic is that substrate used selectively by the host microorganisms that confer health benefits (Santiago et al., 2017). The combination of probiotic and prebiotic agents forms a symbiotic product, which can improve the survival of probiotic bacteria in the gastrointestinal tract, increasing its effect (Mugambi et al., 2014). Therefore, the proposal of the consumption of functional foods with the addition of probiotics and prebiotics can contribute to the improvement or prevention of gastrointestinal diseases, obesity, among others (De Morais, 2016). This document aims to analyze the relationship between the microbiota and the consumption of symbiotic foods, and its benefits for the consumer and the food industry, seeking to highlight its technological properties for the development of functional food.

Diseases Related to the Digestive System

The digestive system is made up of the mouth, pharynx, esophagus, stomach, small intestine, large intestine, and anus. Through chemical and mechanical processes, the digestive system is capable of digesting food until the necessary nutrients are absorbed and transported to the cells (Ascencio, 2012). Among the functions of the digestive system are ingestion, secretion, mixing and propulsion, digestion, absorption, and defecation. The correct functioning of the digestive system will provide the best use of nutrients, and overall health for the body (Mahan et al., 2013).

Today, wrong eating habits, sedentary life, and lack of sleep prevent the adequate assimilation of nutrients, which leads to the development of chronic diseases (Soca, 2017). On the other hand, inadequate digestion, as well as, an unbalanced or incomplete diet, lead to serious diseases. Chronic inflammation, obesity, irritable bowel syndrome, chronic nonspecific ulcerative colitis, diverticulosis, Crohn's disease, autoimmune diseases, gastroenteritis, and cancer are some examples (Ukleja, 2017). The role that intestinal health plays in the origin of these diseases begins with the food that is consumed. Therefore, different investigations have focused on the search for compounds that benefit the functionality of the digestive system (Olveira and González, 2016). There is evidence to suggest that several of the gastrointestinal diseases are associated with alterations in the intestinal microbiome, specifically irritable bowel syndrome, obesity, intolerances, inflammatory bowel disease, and bacterial overgrowth of the small intestine (Chaluvadi et al., 2016). Therefore, the consumption of symbiotic foods, with probiotics and prebiotics, can improve the prognosis of the mentioned diseases (Putta et al., 2018).

Obesity

Obesity is a disease where excess adiposity results in an increase in risks for the human health, this adiposity accumulates in different compartments of the body, causing a mechanical understanding of intra-abdominal organs, besides, this disease has a close relationship with the gastrointestinal system (Calderón and Acosta, 2019). The prevalence of obesity has increased worldwide in recent years, affecting in 2016 one out of three adults worldwide, with 650 million people being obese, this represents 13% of the adult population (WHO, 2020). In countries like Mexico, obesity is considered a pandemic, given that it is one of the two countries with the highest prevalence of obesity. In 2018, it was reported that 75.2% of the Mexican population is affected by overweight and obesity, 36.1% corresponding to people with obesity (ENSANUT, 2018). These high percentages are linked to the risk of developing diabetes, hypertension, and dyslipidemia, affecting the quality of life and productivity of people who suffer from them. Generating high costs for the health system and premature mortality (Campos et al., 2018). In this sense, in Mexico, it has been diagnosed that 8.6 million people over 20 years old have diabetes mellitus (ENSANUT, 2018).

Obesity causes a low-grade systemic inflammatory state, because the adipocytes near the blood vessels receive little blood flow, facilitating hypoxia of the adipose tissue. These inflammatory changes contribute to the activation of inflammatory cells, and an increase in the secretion of proinflammatory cytokines (Heymsfield, 2017), inducing a state of insulin resistance, which favors the development of type 2 diabetes mellitus (León-Pedroza, 2015). The brain-intestine axis is a key part of the development of this disease because the process of food intake is mainly regulated by hormones synthesized in the digestive system, and neurotransmitters and neural centers located in the hypothalamus that determines food intake and energy expenditure (Grill and Hayes, 2012). In the hypothalamus, there is an arcuate nucleus, which has two types of neurons that express different peptides. One group of neurons expresses orexyienic peptides, while the other group expresses anorexigenic peptides (Austin, 2009). Their interaction is crucial for balanced caloric intake. Because the blood-brain barrier is semipermeable, and it is exposed to the action of different hormones, mainly those from the intestine. The gastrointestinal tract, through different mechanisms, provides the necessary signals for the hypothalamus to modulate the amount of food eaten, and its availability for metabolism (Clemmensen, 2017).

There are two hormones (leptin and ghrelin) related to obesity. Leptin is released by adipose tissue; its levels are correlated to those of body fat. This hormone provides a signal that inhibits the activity of neurons in the hypothalamus that synthesize orexigenic peptides and stimulates those that produce anorexigenic peptides, leading to decreased food intake and increased energy expenditure (Waterson and Horvath, 2015). Ghrelin is produced mainly by the cells of the stomach and its levels increase before a meal, within its effects is: increase of food intake, causing an orexigenic response. In contrast to leptin, ghrelin stimulates the orexigenic neurons of the hypothalamus (Mishra, 2016). In obese people, it has been shown that the intestinal microbiome is different from that of a thin person, as a result, a change in the metabolism of intestinal bacteria is generated, decreasing the synthesis of short-chain fatty acids, such as acetate, butyrate, and propionate. Each of these specific functions in the human body affects different mechanisms, that induce the anti-inflammatory process and a hormonal imbalance in obesity (Tolhurst et al., 2012; Sperry et al., 2018; Sun et al., 2019).

Treatment for obesity includes the collaboration of a multidisciplinary team of doctors, nutritionists, psychologists, and exercise specialists, where modifications in food consumption and physical activity are mainly recommended (WHO, 2020). Emphasis is placed on carrying out a caloric deficit diet, with an increase in the consumption of fiber from fruits, vegetables, whole grains, and legumes, as well as, exercising at least 150 minutes per week with moderate intensity, and always keep moving (May Clinic, 2020). In some situations, the doctor may choose to use medications as an aid in treatment. However, to resort to this strategy, several aspects must be considered to be a candidate. Qsymia, Contrave, Lorcaserin, and Orlistat are medications used frequently (Taylor, 2019).

Irritable Bowel Syndrome

Irritable bowel syndrome (IBS) is a chronic functional disorder at the gastrointestinal level, where there are a series of symptoms that range from abdominal pain, bloating, bloating, and periods with the presence of constipation and diarrhea. These are triggered by food, stress, and hormones (Defrees and Bailey, 2017). Prevalence of 10-15% has been reported in the world population in people aged 18 to 34 years (Farmer and Ruffle, 2019). In Mexico, there is a prevalence of 4.4-35% under the Rome III Diagnosis criteria, while the Ministry of Health reports data ranging between 16-30% (Carmona et al., 2016; Ministry of Health, 2016). Having a higher occurrence in women, affecting the quality of life of the patient and their mental state, causing absenteeism (Ministry of Health, 2016). The causes of this syndrome are multifactorial, among which are: alterations in the autonomic nervous system, which cause changes in intestinal motility. Currently, it is known that episodes suffering from gastroenteritis can lead to the development of irritable bowel syndrome (Lacy, 2015). This type of patients come to present modifications in the intestinal microbiome, since there is an increase in the number of Firmicutes class bacteria, concerning a reduction in Bacteroides (Alammar and Stein, 2019).

Effective treatment for IBS is a dietary intervention, which includes changes in food consumption. Different types of diets can be administered depending on the patient's case, among which are; a FODMAP diet, low in fermentable carbohydrates such as disaccharides, monosaccharides, and polyols, present in foods such as wheat, garlic, onion, legumes, dairy products, honey, apples, pears and artificial sweeteners (Lacy, 2015; Eswaran et al., 2017). In addition, nutritional treatment includes a low fiber diet with a gradual increase in it, gluten-free, and lactose-free (McRorie, 2019). Besides, it has been mentioned that the consumption of probiotics, as well as a diet low in fermentable carbohydrates and high in omega 3, helps to remit the disease (Mokhtari and Hekmatdoost, 2019) since it improves the imbalance of the microbiota and in turn the patient's health (Uranga et al., 2016). Among the medications used to treat IBS are antispasmodics, antidiarrheals, and laxatives (Farmer and Ruffle, 2019).

Allergies and Intolerances

The great variety of products offered by the food industry has led to the appearance of adverse reactions to some of these foods, such as food allergies and intolerances (Ruiz et al., 2018). The prevalence of these diseases exceeds 3.6% of the world population, with a higher prevalence in Asian women (Acker, 2017). Lactose intolerance occurs more frequently, and at least 15% of the Mexican adult population suffers from this condition (Rosado, 2016), while food allergy does not exceed 2-3% of the world population and has a prevalence of 8-12% in Mexico (Kleine et al., 2016; Secretaría de Salud, 2019).

Adverse reactions to food are mainly caused by an immune mechanism known as food allergy, instead, when the immune system is not involved, it is known as food intolerance (Kleine et al., 2016). Food allergy is mediated by Immunoglobulins E (IgE). In addition to being part of the most common inflammatory problems, it causes respiratory, gastrointestinal, and skin problems and can even lead to anaphylactic shock (Foong, 2017). This disease is highly related to the composition of the microbiota, because patients with allergies have a lower number of bacteria of the *Ruminococceceae* (Firmicutes and Clostridia) genus, compared to healthy individuals (Sánchez and Santos, 2018). Likewise, it has been determined that the use of probiotics increases the levels of production of Immunoglobulin A (IgA), improving allergic symptoms, and decreases IgE levels (Park et al., 2018). Thus, improving the presence of certain types of commensal microorganisms in the intestine and their function in barrier epithelia, such as bifidobacteria and lactobacilli (Dzidic, 2017).

Food intolerance is a set of non-immunological adverse reactions to food, avoiding the consumption of the foods that cause it, can solve the condition. They can include disorders such as enzyme deficiency, non-celiac gluten sensitivity, intolerance to additives, among others (Martin, 2019). Its manifestations are mainly composed of excessive gas production, abdominal pain, diarrhea, and headaches depending on the causal mechanism of intolerance. Currently, there is no treatment for these diseases, so it is recommended that patients avoid eating foods that contain the ingredients that cause the condition (Reig-Otero et al., 2017). In most cases of people with this problem, they have nutrient-deficient diets, constant episodes of intestinal inflammation, and the intestinal microbiota is affected (Jiménez et al., 2016).

Inflammatory Bowel Disease

Inflammatory bowel disease (IBD) includes two conditions, Crohn's disease (CD) and chronic nonspecific ulcerative colitis (UC) (Limketkai et al., 2018). IBD affects 2 million individuals in North America and 3.2 million in Europe (Ananthakrishnan et al., 2020). Regarding Mexico, the prevalence of cases observed in 2015 was 16.5 for CD, and 54.6 for UC for every 100,000 inhabitants (Yamamoto et al., 2020). Ulcerative colitis is a chronic inflammatory disorder of the rectum mucosa, and sometimes the colon with remission and continuous episodes, where it is believed that there is a relationship between genetic factors, the microbiota, and environmental factors (Yamamoto et al., 2018). Complications include toxic megacolon, perforation of the colon, and increased probability of colon cancer (Forbes et al., 2017). It is currently known that diet is a key component that influences the pathogenesis of this disease since the inflammatory response can modify the health of the microbiome (Mokhtari and Hekmatdoost, 2019). In contrast, Crohn's disease is characterized by chronic idiopathic inflammation of the digestive tract, causing opposition to antigen recognition in the intestinal lumen by the immune system. This will trigger a chronic progressive inflammatory response, which in the long term can develop fistulas, ulcers, intestinal obstruction, and colon cancer (Mayo Clinic, 2020). Treatment encompasses the use of anti-inflammatory drugs such as steroids and immunosuppressants, biological therapy, antibiotics, and supplements (Yamamoto et al., 2017).

There are investigations where the possible therapeutic effects of nutrients, and diet are observed in the intervention of inflammatory bowel disease, showing that a balanced microbiome improves the symptoms of this disease (Limketkai et al., 2018). It is suggested that the consumption of probiotics and prebiotics, can improve the symptoms of UC or CD, in conjunction with an adequate diet and the use of medications (Shen et al., 2014). It is also known that a low fermentable carbohydrate diet can improve symptoms but contributes to poor nutrition, and dysbiosis at the intestinal level in the patient (Prince et al., 2016).

Bacterial Overgrowth of the Small Intestine

Bacterial overgrowth of the small intestine (SIBO) is defined as the presence of a greater number of bacteria (greater than 105 colony-forming units/mL), in the proximal small intestine (Shah and Holtmann, 2019). In this part of the intestine, there are conditions that prevent bacterial overgrowth, some of these conditions are gastric acid, intestinal motility, the ileocecal valve, and the production of immunoglobulins A (IgA), pancreatic juice, and bile (Bohm et al., 2013). Dysbiosis of the intestinal microbiota (alteration in the composition, number, and function of microorganisms) is associated with the pathophysiology of intestinal and extraintestinal disorders (Krajicek et al., 2016). The etiology of SIBO is associated with alterations of the previously mentioned mechanisms. It is also known that microbiota dysbiosis in pathologies such as irritable bowel syndrome, celiac disease, and obesity, is a crucial factor for the development of inflammation, which is why they are linked to SIBO (Zheng et al., 2015; Alammar and Stein; 2019). Its symptoms include abdominal pain and discomfort, diarrhea, steatorrhea, weight loss, and poor nutrient absorption due to structural changes in the intestinal villi (Adike and DiBaise, 2018). Its treatment includes the use of antibiotics, probiotics, and prebiotics, to improve the intestinal barrier, reduce the pH and thus inhibit bacterial adhesion to enterocytes, and at the same time nourish beneficial bacterial strains (Bohm et al., 2013).

Symbiotics

The relationship between the intestinal microbiota and the host cells leads to benefits on the immune system, resistance and inhibition of the translocation of pathogenic microorganisms, and regulation of metabolism (Conlon and Bird, 2015). Bacterial metabolism plays a fundamental role in the intestinal environment, with lifestyle being one of the factors that affect the composition of the intestinal microbiota (Singh et al., 2017). Among the functions of the microbiome, fiber digestion and the secretion of essential metabolites such as amino acids, short and medium-chain fatty acids, neurotransmitters, vitamins, among others, can be mentioned (Putta et al., 2018).

The definition of symbiotic emphasizes products that contain probiotics and prebiotics, where there is a relationship of mutual benefit, and a synergistic effect is generated (Acker and Cash, 2017). In this case, the prebiotic component favors the survival of the probiotic, achieving that it supports variations in the levels of oxygen, pH, and temperature of the gastrointestinal tract (Markowiak and Slizewska, 2017). Therefore, symbiotic functional foods are those that contain prebiotics and probiotics, potentiating their effect, improving the health benefits for the host (Beltrán Vaquero et al., 2015). Table 1 lists some examples of symbiotics. The symbiotics most used in food are made up of *Lactobacillus* and *Bifidobacterium* strains, on the other hand, the most widely used prebiotics are inulin or galactooligosaccharides (Mugambi et al., 2014).

Table 1. Main symbiotics used in clinical studies

Formulation	Probiotic	Prebiotic	Reference
1.Combination (1)	*Lactobacillus plantarum 299*	10 g oat fiber	World Gastroenterology Organization Practice Guideline, 2017.
2.Combination (2)	Mix of *Lactobacillus paracasei (Lpc37), L. rhamnosus (HN001), L. acidophilus (NCFM), Bifidobacterium lactis (HN019)*	6 g Fructo-oligosaccharides	Waitzberg et al., 2013.
3.Combination (3)	*Bacillus coagulans*	Fructo-oligosaccharides	Rogha et al., 2014.
4. Combination (4)	*Lactobacillus sporogens*	Fructo-oligosaccharides	World Gastroenterology Organization Practice Guideline, 2017.

Table 1. (Continued)

Formulation	Probiotic	Prebiotic	Reference
5. Synbiotic 2000 Forte (5)	101 CFU of each: *Pediococcus pentoseceus 5-33:3, Leuconostoc mesenteroides 32-77:1, Lactobacillus paracasei sp. paracasei 19, Lactobacillus plantarum 2362*	2,5 g of each of: inulin, oat fiber, pectin and resistant starch	World Gastroenterology Organization Practice Guideline, 2017.
6. Combination (6)	*Lactobacillus rhamnosus GG, Bifidobacterium lactis Bb12*	Oligofructose + inulin (SYN1)	World Gastroenterology Organization Practice Guideline, 2017.
7. Golden Bifid (7)	*Bifidobacterium bifidum, Lactobacillus bulgaricus, Streptococcus thermophilus*	Fructo-oligosaccharides	World Gastroenterology Organization Practice Guideline, 2017.
8. Combination (8)	*Lactobacillus reuteri DSM 17938*	Lactulose	Ojetti et al., 2014; Schumann 2002.

In humans, the consumption of symbiotics has positively modified the composition of the microbiota in patients with metabolic, gastrointestinal, and infection pathologies (Figure 1). These modifications lead to intestinal immune modulation and the migration of immune cells to the peripheral blood. The administration of symbiotics and their beneficial effects depends on the type of pathology for which they are supplied, and the eating habits of the consumer (Jirillo et al., 2016).

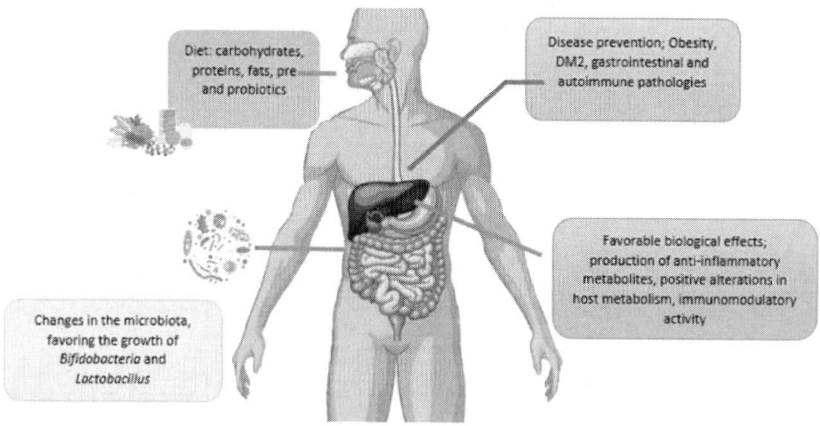

Figure 1. Health impact of the consumption of probiotics and prebiotics.

The formulation of the probiotics and prebiotics mixtures to develop a symbiotic food requires a long evaluation process. Therefore, probiotic and prebiotic combinations are tested to determine which has the best synergy *in vitro* and *in vivo studies, combinations of in vivo* tests, both in animal models and human studies allow evaluating its effectiveness. *In vitro* studies simulate the conditions of the digestive tract, expecting probiotics to reach the intestine almost intact, while prebiotic substances should not be broken down or absorbed to promote a symbiotic effect in the colon (Hwang and Lee, 2019). The correct way to evaluate the functionality and synergistic characteristics of the symbiotic combination is to perform tests *in vitro,* and in animal models, to then be tested in humans. This type of clinical trial must follow the test protocol ethically acceptable by the corresponding committee, and the survival and reproduction of the probiotic component must be demonstrated, as well as, the presence of the prebiotic substance (Rajam and Anandharamakrishnan, 2015).

A symbiotic food can naturally contain within its formulation, a probiotic or a prebiotic component, being the case of fermented milk drinks, where the prebiotic factor is added to look for a symbiosis relationship. On the other hand, in the case of foods whose matrix is mainly made up of ingredients of a prebiotic nature, they can be supplemented with probiotics to obtain the same symbiosis effect (Hwang and Lee, 2019). Another application of symbiotics is during the preparation of coatings with prebiotics, to increase the viability of the probiotic in the digestive tract (Bersaneti et al., 2019). It should be noted that the effectiveness of prebiotics differs between the different genera and species of microorganisms. The degradation and use of prebiotics by microorganisms require that they possess and express specific genes for the hydrolysis and transport of prebiotic substances (Li et al., 2020).

Effects of Symbiotics on Diseases

The administration of symbiotics has been tested in patients undergoing surgical procedures for colon resection; they can attenuate the inflammatory state, stimulate the intestinal function, and reduce complications, in addition to reducing the use of antibiotics and the hospital stay. The effect on inflammation is evidenced due to the decrease in inflammatory markers such as C-reactive protein (CRP) and plasma cytokine IL-6 (Polakowski et al., 2018). Likewise, the use of symbiotics in patients with colonic polyps and colon cancer reduced the theoretical risk of colon cancer, by presenting an improvement in fecal flora and different genetic, cellular, inflammatory, and immunological biomarkers. While in IBD an effect on the colonization of

Lactobacillus, and endogenous Bifidobacteria and inflammatory parameters was observed, favoring the production of short-chain fatty acids (SCFA), and avoiding the appearance of the disease (Olveira and González, 2016). On the other hand, effects on body mass index and waist circumference have been found in obese patients, decreasing their values. Attenuating the levels of pro-inflammatory markers such as lipoprotein-associated with phospholipase A2 (Lp-PLA2), and low-density lipoprotein (LDL), which significantly reduces the inflammatory and oxidative profile in the participants (Cao et al., 2020; Jung et al., 2015). At the same time, bioactive metabolites produced by probiotic bacteria were shown to have an impact on caloric intake homeostasis, due to its agonist effect on intestinal hormones and receptors (Byrne et al., 2015). Figure 2 shows the effect of consuming symbiotics on health.

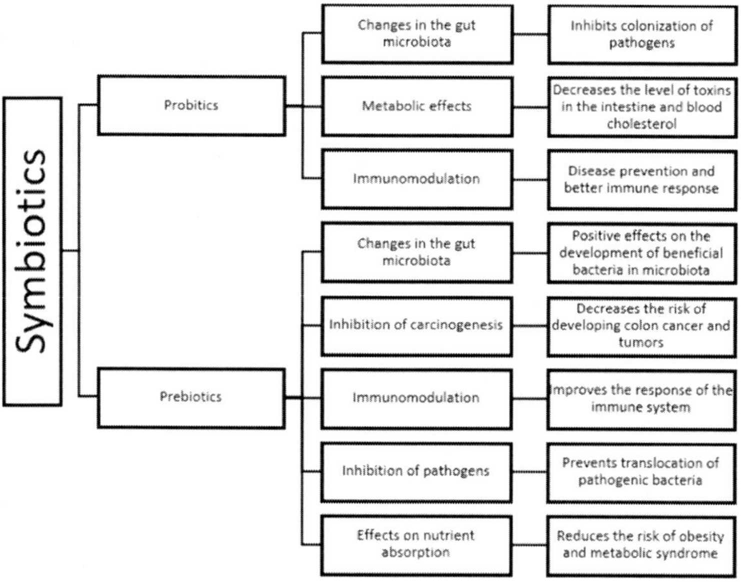

Figure 2. Mechanism of action of symbiotics and their effects on health.

Probiotics

The term probiotic was first introduced by Lilly and Stillwell in 1965, defining it as that factor of microbiological origin that stimulates the growth of other organisms (Villanueva, 2015). Probiotics are defined by the Food and

Agriculture Organization of the United Nations (FAO) and the World Health Organization as living organisms, which, when administered to a host in adequate amounts, confer health benefits (FAO/WHO, 2001). Probiotics generally belong to the group of lactic acid bacteria and comprise the aerobic and anaerobic facultative groups, the majority being Gram-positive bacteria (Salminen and Salminen, 1997), the main species of probiotics that are integrated into food belong to the genera: *Lactobacillus, Bifidobacterium* and *Termophilus*. Non-bacterial microorganisms such as *Saccharomyces boulardii* (Román, 2013) are also used. Furthermore, probiotics provide many health benefits, since they improve the intestinal microbiota, regulate the number of microorganisms, and stimulate the immune and non-immune mechanisms of the mucosa (Valdovinos et al., 2017).

Mechanism of Action of Probiotics

There are different mechanisms of interaction of probiotics with cells of the immune system. In the case of lactic acid bacteria, it has been observed that they can be taken up by the M cells of the intestinal epithelium, which allows the uptake and transport of antigens from the intestinal lumen to the submucosa and facilitate the stimulation of lymphoid tissue (Nova, 2007). They also cause changes in the gastrointestinal environment, such as the induction of an acidic pH below 4, by the production of SCFA, such as acetate and butyrate (Agudelo, 2016). In addition, they have the ability to adhere to enterocytes and colon cells, affecting the composition of the intestinal ecosystem, by increasing the barrier effect not dependent on the immune system and the production of bacteriocins, thus preventing the passage of pathogenic bacteria to the circulation (Gusils, 2002; Tormo, 2006; Alavi et al., 2012).

The microbiota microorganisms act by stimulating the production of mucin; inhibiting pathogens, causing a decrease in intestinal permeability, activating macrophages, and increasing phagocytic capacity (Sánchez and Santo, 2018). In addition to causing an increase in the production of Immunoglobulins A, M and G, and other regulatory elements (Jirillo et al., 2016). On the other hand, its immunomodulatory effect is due to the fact that there is a large surface area of lymphoid tissues within the intestinal mucosa, and probiotic bacteria can be captured by M cells present in the epithelium, stimulating lymphoid tissue. Influencing the response of T cells, responsible for humoral and cellular immunity, destruction of pathogenic microorganisms, and malignant cells (Lin and Zhang, 2017). The effects of probiotics depend largely on the interaction that they have with dendritic cells, and the type of probiotic strain administered and have been observing cases of inhibition or stimulation of the production of IL-10 and IL-12 cytokines, favoring a tolerance or humoral immune response, respectively (Jirillo et al., 2016).

Probiotic Classes

To be able to consider and use a microorganism as a probiotic, it must present safety, functional and technological characteristics (Guarner et al., 2015):

a) They must be non-pathogenic, non-toxic, and not carrying transmissible antibiotic resistance genes, in addition to being isolated from healthy humans.
b) They must survive the physiological conditions of the gastrointestinal tract and show an adhesion to the epithelial surfaces generating an antagonistic activity against pathogens.
c) Have an adequate number of viable strains that lead to the beneficial effect, viability during processing, stability in the product, and during storage, and have scientific evidence of its effectiveness in humans.

To make use of these microorganisms, the type of food matrix and the synergistic potential must be considered. One aspect to consider in the development of products with the addition of probiotics is sensory acceptability, and the viability of the microorganism within the matrix (Serna-Cock, 2014). Most of the species and genera used are generally recognized as safe (GRAS) and are approved by the Food and Drug Administration (FDA) and the European Union (Espitia et al., 2016). The different categories of probiotics are as follows: Medications, foods fermented with probiotics, foods added with probiotics, non-oral probiotics, probiotic animal feed, microbial consortia, probiotic dietary supplement, and probiotic infant formula (Hill et al., 2014; Gibson et al., 2017).

Probiotic drugs are designed to be used in different pathologies, guaranteeing a high concentration of microbial cultures in the form of capsules, powders, or solutions. Fermented foods such as kombucha, kimchi, sauerkraut, and other foods to which probiotics are added, lack evidence of the total number of living microorganisms that they contribute, therefore, they need to be subjected to an evaluation where the number is determined the total bacteria present in the food, to guarantee its probiotic activity (Doğan et al.,

2019). Table 2 lists some examples of the probiotics most used in clinical practice.

Table 2. Main single-strain probiotics used in clinical studies

Lactobacillus	**Bifidobacterium**	**Other bacterias**	**Yeast**
Lactobacillus rhamnosus GG (a)	Bifidobacterium bifidum (f)	Enterococcus faecalis (c)	Saccharomyces boulardii (c)
Lactobacillus plantarum (b)	Bifidobacterium lactis (g)	Escherichia coli Nissle 1917 (serotype 06:K5:H1) (c)	
Lactobacillus sporogens (c)	Bifidobacterium longum (c)	Streptococcus faecalis (h)	
Lactobacillus reuteri (d)	Bifidobacterium infantis (e)	Streptococcus thermophilus (i)	
Lactobacillus casei (L) (c)	Bifidobacterium animalis (e)	Enterococcus faecium (j)	
Lactobacillus acidophilus LA1 (c)	Bifidobacterium thermophilum (e)		
Lactobacillus paracasei (c)			
Lactobacillus delbrueckii (e)			
Lactobacillus brevis (e)			
Lactobacillus lactis (e)			

a. Dand et al., 2014; b. Ducrotté et al., 2012; c. World Gastroenterology Organisation Practice Guideline, 2017; d. Cimperman et al., 2011; e. Tripahi and Giri, 2014; f. Koning et al., 2008; g. Nabavi et al., 2014; h. Du et al., 2012; i. Hempel et al., 2012; j. Koning et al., 2008.

The beneficial effects of probiotics depend on the dose ingested, so a daily intake of 6-9 CFU/g is recommended. Accordingly, a minimum concentration is added to the juices (Gu and Roberts, 2019). These concentrations are suggested to compensate for the loss of probiotic microorganisms by passage through the gastrointestinal tract (Gibson et al., 2017). Table 3 mentions some of the most commercialized brands of probiotics in the world.

Table 3. Examples of commercial products that have probiotics added

Microorganism	**Comercial Name**	**Company and location**
Bifidobacterium animalis DN 173 010	Activia	Danone/Dannon, EE.UU.
Bifidobacterium breve Yakult	Bifiene	Yakult, Japan
Bifidobacterium infantis 35624	Align	Procter y Gamble, EE.UU.
Bifidobacterium lactis HN019	Howaru Bifido	Danisco, Denmark
Enterococcus LAB SF 68	Bioflorin	Cerbios-Pharma, Switzerland
Lactobacillus casei DN-114 001	Actimel	Danone/Dannon, EE.UU.

Table 3. (Continued)

Microorganism	Comercial Name	Company and location
Lactobacillus casei F19	Culturelle	Arla Foods, Denmark
Lactobacillus casei Shirota	Yakult	Yakult, Japan
Lactobacillus rhamnosus LB21	Verum	Norrmejerier, Switzerland
Lactobacillus reuteri DSM 17938	L. reuteri Protectis	BioGaia, Switzerland
Bacillus clausii cepas O/C, NR, SIN and *T*	Enterogermina	Sanofi-Aventis, France
Lactobacillus acidophilus CL1285 and *Lactobacillus casei Lbc80r*	Bio K+	Bio K+ International, Canada
Bifidobacterium animalis subesp. lactis Bb-12	Chr. Hansen	Chr. Hansen, Denmark
Escherichia coli Nissle 1917	Mutaflor	Ardeypharm, Germany
Lactobacillus helveticus R0052 and *L. rhamnosus R0011*	A'Biotica and others	Institut Rosell, Canada

Source: World Gastroenterology Organization Practice Guideline, 2017.

Probiotics in Food

Recently, the interest in the use of probiotics in food has increased, due to the beneficial effects conferred on them. This trend has led to the development of products in the food industry, which in one way or another incorporate probiotics into their formulation (De Morais, 2016). The minimum concentration of a probiotic in food is 10^6 CFU/g, to be considered as a functional food (Guimaraes et al., 2019).

In the food industry, probiotics are found mainly in fermented and non-fermented dairy products, such as fruits, vegetables, and dried meat. They can also be microencapsulated to be added to unfermented foods so that they can withstand the conditions of the food matrix and maintain their viability when passing through the digestive tract (Espitia et al., 2016). Microencapsulation is mainly intended to protect against factors such as pH, oxygen, light, among others, reducing the reactivity of probiotics to the environment to avoid their degradation, in the same way, unpleasant tastes and odors are hidden (Nguyen et al., 2016). In addition to the fact that the technological processes used can also be harmful to these microorganisms, damaging the structure of the probiotic microorganism, due to a decrease in the amount of water (Romano et al., 2016). Among the most widely used methods for encapsulation, spray drying, electrospraying, extrusion, emulsion, and coacervation are mentioned (De Prisco and Mauriello, 2016).

Probiotics are also used for the fermentation of products, to obtain, typical organoleptic characteristics of the food. In this process, the texture of the food is modified, this depends on the type of bacteria and the process used (Blana et al., 2014). Similarly, probiotics have been incorporated into coatings and packaging for their antimicrobial properties, to extend shelf life and maintain food stability. This is known as active packaging, which is an alternative to avoid the use of antibiotics, replacing them with probiotics, and releasing active compounds to preserve food, acting as a barrier against oxygen and reducing humidity and gas transfer rates (Vieira da Silva et al., 2016). Edible coatings added with probiotics have been used in products such as bread, cookies, and fruit (Soukoulis et al., 2014).

Effects of Probiotics on Diseases

There are different reports on the beneficial effects of probiotics in various disease states. The microbiota is known to be different in obese and thin people, supplementation with probiotics can help modulate the gut microbiome, depending on the species and strain used. The use of probiotics as an adjuvant therapy has been shown to reduce body mass index and waist circumference in the control of obesity and its comorbidities (Calderón and Acosta, 2019). The mechanisms that explain these benefits are attributed to the reduction of the inflammatory and oxidative profile (Jung et al., 2015) and the metabolites secreted by probiotic microorganisms, mainly the SCFA resulting from the fermentation of non-digestible compounds (Byrne et al., 2015). Another effect of probiotics is on irritable bowel syndrome (IBS), where it has been observed that some specific strains can alleviate symptoms such as abdominal pain, bloating or bloating, constipation, and bowel habits in this type of patient (Hungin et al., 2018).

In food allergies and intolerances, the beneficial effects of probiotics in the patient have also been documented, as is the case of people with allergies in which the consumption of probiotics increases IgA, which improves allergy symptoms (Sánchez - Salguero et al., 2018). While probiotics help regulate the number of microorganisms in the microbiota in patients with dysbiosis due to food intolerances (Sharma and Bhagatwala, 2019). As for IBD, the use of multi-strains has shown remission of UC. However, the consumption of probiotics in CD has not been shown to have satisfactory results in the remission of the disease (Floch et al., 2015; Guslandi et al., 2015).

International Regulations

Nowadays, the use of supplements based on probiotics has had exponential growth, being available to the entire population, making indiscriminate use of the different strains that the supplements contain, without considering that each strain and dose have a specific function within the human body (Neffe et al., 2018). In some countries, such as Mexico, its use is mainly regulated by official regulations based on Codex Alimentarius documents, while in the United States, they are regulated by the FDA, incorporating them in the category of drugs and supplements. Probiotic products are regulated based on the product category in which they are cataloged and are classified as food or food additives (Hoffmann et al., 2012; US Food and Drug Administration, 2016). In European legislation, the term probiotic is not regulated, since it implies a beneficial effect on health, and its health effect must be declared within its labeling. In addition to that in Europe, they are considered as new foods, those that contain microorganisms that have not had a traditional use in the production of food, being regulated under regulation (EC) No. 2015/2283 (European Commission, 2015). It is worth mentioning that the declaration of healthy properties a food is regulated under Regulation (EC) no 2006/1924 (European Commission, 2006), going through a thorough safety evaluation before its commercialization (Espitia et al., 2016). In contrast, in Brazil, the regulation of products with bioactive substances, and isolates of probiotics with functional or health properties (RDC No 2/2002 ANVISA), is used as a guide for the safety, registration, and marketing of this type of product. Being very similar to the regulations handled by the FDA in the United States (ANVISA, 2002).

Prebiotics

The importance of prebiotics in gastrointestinal health has recently increased, these compounds are increasingly being used to maintain the microbiome health or to restore bacterial balance (Paulo et al., 2019). They can also be added to different food formulations, for the development of products with functional properties (Hutkins et al., 2016). Prebiotics are nondigestible food ingredients that, when metabolized by some genera of microbiota bacteria, provide a beneficial effect on the host, selectively stimulating the growth of bifidobacteria and *Lactobacillus,* in addition to controlling the growth of pathogenic bacteria in the colon (Santiago et al., 2017). The main prebiotics are identified as non-digestible carbohydrates such as polysaccharides (resistant starch, pectin, and dextrin), and oligosaccharides such as fructooligosaccharides (FOS), lactulose, inulin, and other compounds (Colanthonium et al., 2019). These compounds are fermented by the microbiota of the intestine, synthesizing some SCFA among them: propionate, butyrate, and acetate. SCFA production is related to intestinal function, mineral absorption, regulation of lipid, and glucose metabolism, in addition to reducing the risk of developing colon cancer (Paulo et al., 2019).

Mechanism of Action

Prebiotics have a selective effect on the microbiota, improving the health of the host (Gibson et al., 2017). For a substance to be considered prebiotic, it must resist the digestion process, considering gastric acid and enzymatic hydrolysis, in addition to being fermented by the microbiota of the intestine, and have an effect on the selective growth of bacteria (Scott et al., 2019). The effects of prebiotics depend on its type and concentration, in addition to the number of bacteria present in the intestine (Clark et al., 2012). The glycosidic bonds of these carbohydrates are resistant to the hydrolysis of digestive enzymes, reaching the colon intact where they are the substrate to be fermented by the microbiota, producing short-chain fatty acids (Davani-Davari et al., 2019). The product obtained from the metabolism of the intestinal microbiome; SCFAss are organic acids of 2 to 5 carbons produced by the fermentation of prebiotics in the intestinal microbiome, including acetate, propionate, butyrate, and valerate. The SCFA are mainly used as an energy source for the monocytes. In addition to being regulators of physiological processes, related to normal intestinal function, they also modify the volume of feces due to an increase in microbial biomass. The production of SCFAs is influenced by the stimulation of prebiotics on the microbiota. An example is inulin, which increases acetate and butyrate levels, while xylooligosaccharides produce a greater amount of acetate (Gyawali et al., 2019).

One of the effects attributed to prebiotics is the decrease of blood cholesterol, because butyrate inhibits cholesterol synthesis, providing an energy source for epithelial cells. On the other hand, propionate inhibits the synthesis of fatty acids in the liver and reduces the secretion of triglycerides (Lin et al., 2012). SCFA are also known to inhibit the growth of tumor cells in the colon, as butyrate absorbed by colon cells, prevents tumor formation by inducing apoptosis, as well as, protecting the body from carcinogens (Markowiak and Slizewska, 2017).

Types of Prebiotics

Currently, prebiotics have been used to improve the health of the adult consumer and pediatric patients who present some pathological condition such as diabetes, cancer, kidney disease, ulcers, immunosuppression, and metabolic stress, integrating them into food or in their consumption (De Paula Farias et al., 2019). Among the most studied prebiotics are polysaccharides and oligosaccharides. Polysaccharides include resistant starch, pectin, and dextrin. Oligosaccharides encompass fructooligosaccharides (FOS), galactooligosaccharides (GOS), xylooligosaccharides (XOS), isomaltooligosaccharides (IMO), raffinosaoligosccharides (RFOS), arabinoxylan oligosaccharides, lactulose, inulin. Some others are nondigestible carbohydrates and are fermentable for the human gut microbiota (Colantonio et al., 2019; Mano et al., 2017; Mohanty et al., 2018).

Table 4. Main prebiotics used in clinical studies

Non-digestible carbohydrates	Oligosaccharides	Compound mixes
Oat fiber	Fructo-oligosaccharides (FOS)	Beneo Synergy 1 (SYN1): oligofructose-inulin
Sprouted barley (rich in hemicellulose)	Galacto-oligosaccharides (GOS)	
Hydrolyzed guar gum	Inulin	
Resistant starch	Lactulose	
Plantago ovata	Isomaltulose	
Pectin		
Betaglucan		

Source: World Gastroenterology Organization Practice Guideline, 2017.

Other authors classify prebiotics into three main groups: Polyols (lactulose, xylitol, mannitol, lactitol), Oligosaccharides (FOS, GOS, XOS, IMO, RFOS, isomaltulose, inulin) and Fibers (Cellulose, dextrins, pectins, beta-glucans) (Mohanty et al., 2018). However, according to the described definition, other compounds such as linoleic acid, polyunsaturated fatty acids, phenolic and phytochemical compounds may also have prebiotic activities (Dávila et al., 2019). Prebiotics can be used as supplements and incorporated into food (Ashwini et al., 2019) (Table 4).

Prebiotics in Food

Among the most studied prebiotics are fructans, which are polymers, which are composed of fructose units, with β (2-1) bonds type called inulins and/or β (2-6) called levans; present in some genera of bacteria, fungi, algae, and plants. When they have several fructose residues between 2 and 10, they are called oligofructose or fructooligosaccharides (FOS) (Ávila, 2014). Agave pineapple has carbohydrates between 13 and 17 g/100 g, a fraction of this is made up of fructans (Santiago et al., 2017). Being a part of the agave used to prepare fructose syrup, and extract fructans from agave. The glycosidic bond that links fructose cannot be hydrolyzed in the digestive tract, considering it as soluble fiber, which upon reaching the colon can be used by some bacteria, generating a prebiotic potential (Zamora et al., 2014). Some effects of the consumption of fructans are the synthesis of vitamins, a lower incidence in digestive disorders, and a decrease in the blood levels of phospholipids and triglycerides and intestinal immunomodulation (Ávila, 2014). It should be noted that the presence of prebiotic molecules must be guaranteed in processed food, otherwise, their prebiotic effect will not be valid (Santiago et al., 2017).

Prebiotics, like oligosaccharides and polyols, are used as low glycemic index sweeteners in the development of functional foods, due to their low energy content (Ávila, 2014). It is worth mentioning that they can be used as fat substitutes, emulsifiers, and texturizers. When using fructans as a fat substitute, the perception of flavor and texture of food can be altered (Santiago *et al.,* 2017), since they imitate the behavior of fat when absorbing water, lubricating and giving a pleasant sensation in the mouth (Zamora *et al.,* 2014). Another type of prebiotics are the polysaccharides, which show a fermentation capacity and a prebiotic effect, among them the resistant starch present in raw potatoes, cooked and refrigerated starch products, and in immature fruits such as bananas. The resistant starch is used in the food industry, incorporating it into several commercial foods due to the properties conferred during processing (Balthazar et al., 2017).

In recent years, the use of prebiotics in food has increased, due to the technological properties that these provide among them: improvements in sensory aspects, texture, and physicochemical characteristics when adding fibers, and used as a replacement for fat and sugars (Guimarães et al., 2019). In the food industry, the substitution of fat for inulin has been carried out in various products, thus affecting rheological properties, depending on the product in which it is applied (Guimarães et al., 2018). An example of this is bakery products, where margarine is replaced by an oligosaccharide, to decrease the caloric intake, and provide a higher fiber content to the mixture, affecting the viscosity and consistency of the product. There is still a wide field of application for the use of prebiotics in food (Paulo et al., 2019).

Effect of Prebiotics on Diseases

Prebiotics have been investigated for their effect on health and the management of different diseases. An example where the activity of prebiotic substances can be evaluated is obesity, due to the production of SCFA synthesized by probiotics. Where butyrate and propionate have stimulating effects on anorexic peptides and help regulate weight, while acetate inhibits weight gain, regardless of the suppression of food intake (Lin et al., 2012). In humans, propionate increases leptin expression and stimulates lipogenesis and glucose absorption (Calderón and Acosta, 2019). While in IBD the use of prebiotics favors the growth of *Lactobacillus* and Bifidobacteria, increasing the production of ACGC, especially butyrate, which prevents the adhesion of pathogenic bacteria to the mucosa, decreases the pH and reduces the appearance of UCS events, especially with inulin and fiber from *Plantago ovata*. Aside from reducing abdominal pain and improving symptoms, it is worth mentioning that its use is not recommended as the only treatment in this pathology (Mokhtari and Hekmatdoost, 2019).

International Regulations

The consumption of prebiotics and their sale is regulated by different rules. In the United States, they are regulated under the Food Drugs and Cosmetics Act (FDCA), considering that any substance added to food is defined as an additive that will have to be approved by the FDA (Kumar et al., 2015). While, in the European Union, the prebiotic components that have been identified after 1997, are considered as novel foods, undergoing a thorough evaluation by the European Commission and the European Food Safety Authority (EFSA). It is worth mentioning that EFSA does not recognize the relationship of the effects between the consumption of dietary fiber and prebiotic, so these components must declare on the labeling of food as dietary fiber, and without implying the benefits they provide for health (Kumar et al., 2020). In Canada, this issue is handled under the regulations of the Food and Drug Regulations (FDR), where before the sale of any product, a safety assessment data package must be presented. On the other hand, in Japan, prebiotics can be declared components with healthy properties of any food category (Kumar et al., 2015). Globally, there is no uniformity among countries for food legislation considered as food supplements, dietary supplements, natural products, or complementary medicines (Dwyer et al., 2018).

Future Trends

Currently, there is an increasing demand for the development of functional products, which has led the food industry to implement strategies for the formulation of symbiotic foods (Neffe et al., 2018; Paulo et al., 2019). There is a long list of foods that have been added with some probiotic, prebiotic substance or both. Current trends are inclined to foods with a favorable matrix for the development of microorganisms, among them, we can mention non-alcoholic beverages, such as beer, juices, dairy products, vegetables, fruits, and pastries (Table 5). The symbiotic complexes most used in food, and with the best potential for synergy between both components, are microorganisms of the *Lactobacillus* and *Bifidobacterium* genera, together with galacto-oligosaccharides and phosphor-oligosaccharides (Jirillo et al., 2016).

Table 5. Symbiotic foods, their composition and effectiveness

Symbiotic food	Probiotic	Prebiotic	Effect	Reference
Semi-hard cheese	*Lactobacillus paracasei*	FOS	Viability of *L. paracasei* in the host	(Langa et al., 2019)
Queso Fiordilatte cheese	*Lactobacillus rhamnosus*	FOS	Viability of *L. rhamnosus* in the food	(Angiolillo et al., 2014).
Tomato	*Lactobacillus acidophilus*	Inulina	Viability of *L. acidophilus* in the food	(Dini Staliano et al., 2015).
Fermented milk	*Lactobacillus acidophilus*	Resistant starch	Viability of *L. acidophilus* in the food	(Batista et al., 2017).
Coconut water	*Lactobacillus rhamnosus*	Inulin	Viability of *L. rhamnosus* in the food	(Segura Badilla et al., 2020)
Soy yogurt	*Lactobacillus delbrueckii* and *Streptococcus thermophilus*	FOS	Hypocholesterolemic effect in mice	(Sengupta et al., 2019).

Despite efforts to develop foods with symbiotic properties, there are limitations such as achieving colonization potential (Johnson and Foster, 2018), nutritional preferences, and metabolite profile of the microbiome (Tramontano et al., 2018), in addition to considering the interactions between the host microbiota and the prebiotic (Cukkemane et al., 2019).

Conclusion

Currently, the prevalence of chronic diseases linked to wrong eating habits caused by an accelerated lifestyle has led to the development of functional products, which can attenuate symptoms or prevent the appearance of such pathologies. The consumption of symbiotic foods has proven to be a useful strategy to modulate the symptoms in diseases such as IBS, IBD, allergies, and intolerances, and contribute to the treatment of obesity. More research is still needed to determine which are the best combinations, between the symbiotic components added in each food, and how they work in different conditions, to offer a better benefit to the host. In addition to that better techniques must be sought for the incorporation of probiotic compounds and prebiotics in solid food matrices, for optimal viability of the symbiotic mixture, ensuring that its benefits are used by the human body, and not only enhance sensory characteristics in food.

Acknowledgments

SNCG appreciates the financial support of the National Council of Science and Technology of Mexico (CONACyT) to carry out their postgraduate studies. This study was funded by the Autonomous University of Coahuila.

Conflict of Interest

The authors declare that they have no conflict of interest.

References

Acker, B. W., & Cash, B. D. (2017). Medicinal Foods for Functional GI Disorders. *Current Gastroenterology Reports* 19(12). doi:10.1007/s11894-017-0601-x.

Acker, W. W., Plasek, J. M., Blumenthal, K. G., Lai, K. H., Topaz, M., Seger, D. L., … Zhou, L. (2017). Prevalence of food allergies and intolerances documented in electronic health records. *Journal of Allergy and Clinical Immunology* 140(6): 1587–1591.e1. doi:10.1016/j.jaci.2017.04.006.

Adike, A., & DiBaise, J. K. (2018). Small Intestinal Bacterial Overgrowth. *Gastroenterology Clinics of North America* 47(1): 193–208. doi:10.1016/j.gtc.2017.09.008.

Alammar, N., & Stein, E. (2019). Irritable Bowel Syndrome. *Medical Clinics of North America* 103(1): 137–152. doi:10.1016/j.mcna.2018.08.006.

Ananthakrishnan, A. N., Kaplan, G. G., & Ng, S. C. (2020). Changing Global Epidemiology of Inflammatory Bowel Diseases—Sustaining Healthcare Delivery into the 21st Century. *Clinical Gastroenterology and Hepatology* 18(6): 1252–1260. doi:10.1016/j.cgh.2020.01.028.

Angiolillo, L., Conte, A., Faccia, M., Zambrini, A. V., & Del Nobile, M. A. (2014). A new method to produce synbiotic Fiordilatte cheese. *Innovative Food Science & Emerging Technologies* 22: 180–187. doi:10.1016/j.ifset.2013.09.010.

Ascencio Peralta C. (2012). Capítulo 2: Ingreso y utilización de los alimentos en el sistema digestivo. *Fisiología de la nutrición.* México. Editorial Mc Graw Hill. [Chapter 2: Intake and use of food in the digestive system. *Nutrition Physiology*]

Ashwini, A., Ramya, H. N., Ramkumar, C., Reddy, K. R., Kulkarni, R. V., Abinaya, V., Naveen S., Raghu, A. V. (2019). Reactive mechanism and the applications of bioactive prebiotics for human health: Review. *Journal of Microbiological Methods* 159: 128-137. doi:10.1016/j.mimet.2019.02.019.

Austin, J., & Marks, D. (2009). Hormonal Regulators of Appetite. *International Journal of Pediatric Endocrinology* 2009: 1–9. doi:10.1155/2009/141753.

Balthazar C. F., Silva H., A., Vieira A. H., Neto R. P. C., Cappato L. P., Coimbra P. T., Moraes J., Andrade M. M., Calada V. M. A., Granato D., Freitas M. Q., Tavares M. I. B., Raíces R. S. L., Silva M. C., Cruz, A. G. (2017). Assessing the effects of different prebiotic dietary oligosaccharides in sheep milk ice cream. *Food Research International* 91: 38–46. doi: 10.1016/j.jff. 2017.06.004.

Batista, A. L. D., Silva, R., Cappato, L. P., Ferreira, M. V. S., Nascimento, K. O., Schmiele, M., … Cruz, A. G. (2017). Developing a symbiotic fermented milk using probiotic bacteria and organic green banana flour. *Journal of Functional Foods* 38: 242–250. doi: 10.1016/j.jff.2017.09.037.

Beltrán Vaquero D.A, Crespo Garzón A. E., Rodriguez Bravo T. C., Garcia Iglesias A. (2015). Adultos Mastitis infecciosa: nueva solución para un viejo problema Ayuntamiento de Madrid. *Nutr Hosp.* 31 (Supl. 1): 89-95. doi:10.3305/nh.2015.31.sup1.8714. [Adults Infectious mastitis: a new solution for an old problem Madrid City Council.]

Bersaneti, G. T., Garcia, S., Mali, S., Pedrine Colabone Celligoi, M. A. (2019). Evaluation of the prebiotic activities of edible starch films with the addition of nystose from Bacillus subtilis natto. *LWT* 116: 108502. doi: 10.1016/j.lwt.2019.108502.

Blana, V. A., Grounta, A., Tassou, C. C., Nychas, G. J. E., & Panagou, E. Z. (2014). Inoculated fermentation of green olives with potential probiotic *Lactobacillus pentosus* and *Lactobacillus plantarum* starter cultures isolated from industrially fermented olives. *Food Microbiology* 38: 208–218. doi:10.1016/j.fm.2013.09.007.

Bohm, M., Siwiec, R. M., Wo, J. M. (2013). Diagnosis and Management of Small Intestinal Bacterial Overgrowth. *Nutrition in Clinical Practice,* 28(3): 289–299. doi: 10.1177/0884533613485882.

Byrne C. S., Chambers E. S., Morrison D. J., Frost G. (2015). The role of short chain fatty acids in appetite regulation and energy homeostasis. *Int J Obes* (Lond) 39(9): 1331-1338. doi: 10.1038/ijo.2015.84.

Calderón G., Acosta A. (2019). *The gastrointestinal system and obesity dietary interventions in gastrointestinal diseases.* Clinical Enteric Neuroscience Translational and Epidemiological Research (C.E.N.T.E.R.) 43-62. Mayo Clinic, Rochester, MN, United States.

Campos Nonato I., Cueva Nasu L., González Castell L. D., Hernández Barrera L., Shamah Levy T., González de Cosío Martínez T., Rivera Dommarco J. A. (2018). *La obesidad en México: Estado de la política pública y recomendaciones para su prevención y control.* Capítulo 1: Epidemiología de la obesidad y sus principales comorbilidades en México. Instituto Nacional de Salud Pública. [*Obesity in Mexico: State of public policy and recommendations for its prevention and control.* Chapter 1: Epidemiology of obesity and its main comorbidities in Mexico]

Cao S., Ryan P. M., Salehisahlabadi A., Abdulazeem H. M., Karam G., Černevičiūtė R., Antuševas A., Rahmani J., Zhang Y. (2020). Effect of probiotic and synbiotic formulations on anthropometrics and adiponectin in overweight and obese participants: A systematic review and meta-analysis of randomized controlled trials, *Journal of King Saud University – Science* 32(2): 1738-1748. doi: 10.1016/j.jksus.2020.01.011.

Carmona-Sánchez R., Icaza-Chávez M. E., Bielsa-Fernández M. V., Gómez-Escudero O., Bosques-Padilla F., Coss-Adame E., Esquivel-Ayanegui F., Flores-Rendón A. R., González-Martínez M. A., Huerta-Iga F., López-Colombo A., Méndez-Gutiérrez T. H., Noblre-Hugo A., Nogueira-de Rojas J. R., Raña-Garibay R. H., Remes-Troche J. M., Roesch-Dietlen F., Schmulson M. J., … Zavala-Solares M. R. (2016). Consenso mexicano sobre el síndrome de intestino irritable. *Revista de Gastroenterología de México* 81(3): 149–167. doi: 10.1016/j.rgmx.2016.01.004. [Mexican consensus on irritable bowel syndrome. *Journal of Gastroenterology of Mexico*]

Chaluvadi S., Hotchkiss A. T., Yam K. L. (2016). Gut microbiota. *Probiotics, Prebiotics, and Synbiotics* 515–523. doi: 10.1016/B978-0-12-802189-7.00036-8.

Cimperman L., Bayless G., Best K., Diligente A., Mordarski B., Oster M., et al. (2011). A randomized, double blind, placebo-controlled pilot study of *Lactobacillus reuteri* ATCC 55730 for the prevention of antibiotic-associated diarrhea in hospitalized adults. *J Clin Gastroenterol* 45(9): 785–9. doi: 10.1097/MCG.0b013e3182166a42.

Clark, M. J., Robien, K., Slavin, J. L. (2012). Effect of prebiotics on biomarkers of colorectal cancer in humans: a systematic review. *Nutrition Reviews* 70(8): 436–443. doi: doi.org/10.1111/j.1753-4887.2012.00495.x.

Clemmensen C., Müller T. D., Woods S. C., Berthoud H. R. R., Seeley R. J., Tschöp M. H. (2017). Gut-brain cross-talk in metabolic control Cell 168(5):758–74. doi: Clemmensen C., Müller T. D., Woods S. C., Berthoud H. R. R., Seeley R. J., Tschöp M. H. (2017). Gut-brain cross-talk in metabolic control *Cell* 168(5): 758–74.

Colantonio, A. G., Werner, S. L., Brown, M. (2019). The effects of prebiotics and substances with prebiotic properties on metabolic and inflammatory biomarkers in individuals with type 2 diabetes mellitus: A systematic review. *Journal of the Academy of Nutrition and Dietetics* 120(4): 587-607.e2 doi: 10.1016/j.jand.2018.12.013.

Conlon, M., Bird, A. (2015). The impact of diet and lifestyle on gut microbiota and human health. *Nutrients* 7(1): 17–44. doi: 10.3390/nu7010017.

Dang Y, Reinhardt JD, Zhou X, Zhang G. (2014). The effect of probiotics supplementation on *Helicobacter pylori* eradication rates and side effects during eradication therapy: a meta-analysis. *PloS One* 9(11): e111030. doi: 10.1371/journal.pone.0111030.

Davani-Davari D., Negahdaripour M., Karimzadeh I., Seifan M., Mohkam M., Masoumi S. J., Berenjian A., Ghasemi Y. (2019). Prebiotics: Definition, types, sources, mechanisms, and clinical applications *Foods* 8(3): 92. doi: 10.3390/foods8030092.

Dávila, I., Gullón, B., Alonso, J. L., Labidi, J., Gullón, P. (2019). Vine shoots as new source for the manufacture of prebiotic oligosaccharides. *Carbohydrate Polymers* 207: 34–43. doi: 10.1016/j.carbpol.2018.11.065.

De Morais, E. C. (2016). Prebiotic addition in dairy products. *Probiotics, Prebiotics, and Symbiotics* 37–46. doi: 10.1016/B978-0-12-802189-7.00003-4.

De Paulo Farias, D., Fernandes de Araújo, F., Neri-Numa, I. A., Pastore, G. M. (2019). Prebiotics: Trends in food, health and technological applications. *Trends in Food Science & Technology* 93: 23-25. doi: 10.1016/j.tifs.2019.09.004.

De Prisco, A., Mauriello, G. (2016). Probiotication of foods: A focus on microencapsulation tool. *Trends in Food Science and Technology* 48: 27–39. doi: 10.1016/j.tifs.2015.11.009.

Defrees, D. N., Bailey, J. (2017). Irritable bowel syndrome. *Primary Care: Clinics in Office Practice* 44(4): 655–671. doi: 10.1016/j.pop.2017.07.009.

Doğan, M., Tekiner, İ. H., Demirkesen Biçak, H. (2019). Probiotics from food products and gastrointestinal health. *Dietary Interventions in Gastrointestinal Diseases* 169–177. doi: 10.1016/B978-0-12-814468-8.00013-2.

Du Y. Q., Su T., Fan J. G., Lu Y. X., Zheng P., Li X. H., et al. (2012). Adjuvant probiotics improve the eradication effect of triple therapy for *Helicobacter pylori* infection. *World J Gastroenterol* 18(43): 6302–7. doi: 10.3748/wjg.v18.i43.6302.

Ducrotté P, Sawant P, Jayanthi V. (2012). Clinical trial: *Lactobacillus plantarum* 299v (DSM 9843) improves symptoms of irritable bowel syndrome. *World J Gastroenterol* 18(30):4012–8. doi: 10.3748/wjg.v18.i30.4012.

Dzidic M., Abrahamsson T. R., Artacho A., Björkstén B., Collado M. C., Mira A., et al. (2017). Aberrant IgA responses to the gut microbiota during infancy precede asthma and allergy development. *J Allergy Clin Immunol* 139(3): 1017-1025. doi: 10.1016/j.jaci.2016.06.047.

Espitia, P. J. P., Batista, R. A., Azeredo, H. M. C., Otoni, C. G. (2016). Probiotics and their potential applications in active edible films and coatings. *Food Research International* 90: 42–52. doi: 10.1016/j.foodres.2016.10.026.

Eswaran, S., Farida, J. P., Green, J., Miller, J. D., Chey, W. D. (2017). Nutrition in the management of gastrointestinal diseases and disorders: the evidence for the low FODMAP diet. *Current Opinion in Pharmacology* 37: 151–157. doi: 10.1016/j.coph.2017.10.008.

Farmer, A. D., Ruffle, J. K. (2019). Irritable bowel syndrome. *Medicine.* 47(6). doi: 10.1016/j.mpmed.2019.03.009.

Foong R. X., du Toit G., Fox A. T. (2017). Asthma, food allergy, and how they relate to each other. *Front Pediatr* 9(5). doi: 10.3389/fped.2017.00089.

Forbes A., Escher J., Hébuterne X., et al. (2017). ESPEN guideline: clinical nutrition in inflammatory bowel disease. *Clin Nutr* 36(2): 321–47. doi: 10.1016/j.clnu.2016.12.027.

Gibson G. R., Hutkins R., Sanders M. E., Prescott S. L., Reimer R. A., Salminen S. J., Scott K., Stanton C., Swanson K. S., Cani P. D., et al. (2017). Expert consensus document: The International Scientific Association for Probiotics and Prebiotics (ISAPP) consensus statement on the definition and scope of prebiotics. *Rev Gastroenterol Hepatol* 14: 491–502. doi:10.1038/nrgastro .2017.75.

Gibson, G. R., Probert, H. M., Loo, J. V., Rastall, R. A., Roberfroid, M. B. (2004). Dietary modulation of the human colonic microbiota: updating the concept of prebiotics. *Nutrition Research Reviews* 17(02): 259. doi: 10.1079/ NRR200479.

Gourbeyre, P., Denery, S., Bodinier, M. (2011). Probiotics, prebiotics, and synbiotics: impact on the gut immune system and allergic reactions. *Journal of Leukocyte Biology* 89(5): 685–695. doi: 10.1189/jlb.1109753.

Grill H. J., Hayes M. R. (2012). Hindbrain neurons as an essential hub in the neuroanatomically distributed control of energy balance. *Cell Metab* 16(3): 296–309. doi: 10.1016/j.cmet.2012.06.015.

Grimm, V., Riedel, C. U. (2016). Manipulation of the microbiota using probiotics. *Advances in Experimental Medicine and Biology* 109–117. doi:10.1007/978-3-319-31248-4_8.

Gu, J., Roberts, K. (2019). Probiotics and prebiotics. *Adult Short Bowel Syndrome* 67–80. doi: 10.1016/B978-0-12-814330-8.00006-8.

Guimarães, J. T., Balthazar, C. F., Silva, R., Esmerino, E. A., Silva, M. C., Sant'Ana, A. S., … Cruz, A. G. (2019). Impact of probiotics and prebiotics on food texture. *Current Opinion in Food Science* 33: 38-44. doi: 10.1016/j.cofs.2019.12.002.

Guimarães, J. T., Silva, E. K., Costa, A. L. R., Cunha, R. L., Freitas, M. Q., Meireles, M. A. A., Cruz, A. G. (2018). Manufacturing a prebiotic whey beverage exploring the influence of degree of inulin polymerization. *Food Hydrocolloids* 77: 787–795. doi: 10.1016/j.foodhyd.2017.11.021.

Gyawali, R., Nwamaioha, N., Fiagbor, R., Zimmerman, T., Newman, R. H., Ibrahim, S. A. (2019). The role of prebiotics in disease prevention and health promotion. *Dietary Interventions in Gastrointestinal Diseases* 151–167. doi: 10.1016/B978-0-12-814468-8.00012-0.

Hempel S., Newberry S. J., Maher A. R., Wang Z., Miles J. N. V., Shanman R., et al. (2012). Probiotics for the prevention and treatment of antibiotic-associated diarrhea: a systematic review and meta-analysis. *JAMA*. 307(18): 1959–69. doi: 10.1001/jama.2012.3507.

Heymsfield S. B., Wadden T. A. (2017). Mechanisms, pathophysiology, and management of obesity. *N Engl J Med* 376(3): 254–66. doi: 10.1056/NEJMra1514009.

Hwang H. W., Lee D. W. (2019). Prebiotics: An overview of current researches and industrial applications. *Food Science and Industry* 52(3). doi: 10.1155/2018/5427978.

Jana, T., Acker, B. W., Cash, B. D. (2020). Probiotics and prebiotics, including fibers and medicinal foods. *Clinical and Basic Neurogastroenterology and Motility* 587–600. doi: 10.1016/B978-0-12-813037-7.00042-X.

Jirillo, F., Jirillo, E., Magrone, T. (2016). Synbiotics and the immune system. *Probiotics, Prebiotics, and Synbiotics* 449–458. doi: 10.1016/B978-0-12-802189-7.00030-7.

Jung S., Lee Y. J., Kim M., Kim M., Kwak J. H., Lee J. W., . . . Lee J. H. (2015). Supplementation with two probiotic strains, *Lactobacillus curvatus* HY7601 and *Lactobacillus plantarum* KY1032, reduced body adiposity and Lp-PLA2 activity in overweight subjects. *Journal of Functional Foods* 19: 744-752. doi: 10.1016/j.jff.2015.10.006.

Kleine-Tebbe J., Waßmann-Otto A., Mönnikes H. (2016). Food allergy and intolerance: distinction, definitions and delimitation. *Bundesgesundheitsblatt Gesundheitsforschung Gesundheitsschutz* 59(6):705-22. doi: 10.1007/

s00103-016-2356-1. [*Federal Health Gazette Health Research Health Protection*]

Koning C. J. M., Jonkers D. M. A. E., Stobberingh E. E., Mulder L., Rombouts F. M., Stockbrügger R. W. (2008). The effect of a multispecies probiotic on the intestinal microbiota and bowel movements in healthy volunteers taking the antibiotic amoxycillin. *Am J Gastroenterol.* 103(1): 178–89.

Krajicek, E. J., Hansel, S. L. (2016). Small intestinal bacterial overgrowth. *Mayo Clinic Proceedings,* 91(12): 1828–1833. doi: 10.1016/j.mayocp.2016.07.025.

Kumar Verma, D., Kareem Niamah, A., Patel, A. R., Thakur, M., Singh Sandhu, K., Chávez-González, M. L., … Noe Aguilar, C. (2020). Chemistry and Microbial Sources of Curdlan with Potential Application and Safety Regulations as Prebiotic in Food and Health. *Food Research International,* 109136. doi: 10.1016/j.foodres.2020.109136.

Lacy, B. E. (2015). The science, evidence, and practice of dietary interventions in irritable bowel syndrome. *Clinical Gastroenterology and Hepatology* 13(11): 1899–1906. doi: 10.1016/j.cgh.2015.02.043.

Langa, S., van den Bulck, E., Peirotén, A., Gaya, P., Schols, H. A., Arqués, J. L. (2019). Application of lactobacilli and prebiotic oligosaccharides for the development of a symbiotic semi-hard cheese. *LWT,* 114: 108361. doi: 10.1016/j.lwt.2019.108361.

León Pedroza J., González Tapia L., del Olmo Gil E., Castellanos Rodríguez D., Escobedo G., González Chávez A. (2015). Low-grade systemic inflammation and the development of metabolic diseases: from the molecular evidence to the clinical practice. *Cir Cir* (Engl Ed) 83(6): 543–51. doi: 10.1016/j.circir.2015.05.041.

Li H., Zhang T., Li C., Zheng S., Li H., Yu J. (2020). Development of a microencapsulated symbiotic product and its application in yoghurt. *LWT* 122 (10): 9033. doi: 10.1016/j.lwt.2020.109033.

Limketkai, B. N., Wolf, A., Parian, A. M. (2018). Nutritional interventions in the patient with inflammatory bowel disease. *Gastroenterology Clinics of North America* 47(1): 155–177. doi: 10.1016/j.gtc.2017.09.007.

Lin, H. V., Frassetto, A., Kowalik Jr, E. J., Nawrocki, A. R., Lu, M. M., Kosinski, J. R., … Marsh, D. J. (2012). Butyrate and propionate protect against diet-induced obesity and regulate gut hormones via free fatty acid receptor 3-independent mechanisms. *PLoS ONE,* 7(4): e35240. doi: 10.1371/journal.pone.0035240.

Lin, L., Zhang, J. (2017). Role of intestinal microbiota and metabolites on gut homeostasis and human diseases. *BMC Immunology* 18(1). doi: 10.1186/s12865-016-0187-3.

Mahan L. K., Escott Stump S., Raymond J. L. (2013). *Krause Dietoterapia.* Editorial Elsevier España. Barcelona, España. [*Krause Diet Therapy.*]

Mano, M. C. R., Neri-Numa, I. A., Silva, J. B. da, Paulino, B. N., Pessoa, M. G., Pastore, G. M. (2017). Oligosaccharide biotechnology: An approach of prebiotic revolution on the industry. *Applied Microbiology and Biotechnology* 102(1): 17–37. doi: 10.1007/s00253-017-8564-2.

Markowiak P., Slizewska K. (2017). Effects of Probiotics, Prebiotics, and Symbiotic on Human Health. *Nutrients* 9(9): 1021. doi: 10.3390/nu 9091021.

Martin, V. M. (2019). Food intolerance and non-celiac gluten sensitivity. Reference Module in Biomedical Sciences. *Encyclopedia of Gastroenterology,* 2nd Edition.

Mayo Clinic (2020). *Enfermedad de Crohn: síntomas y causas.* https://www.mayoclinic.org/es-es/diseases-conditions/crohns-disease/symptoms-causes/syc-20353304 (Accessed 26 March 2020). [*Crohn's disease: symptoms and causes*]

Mayo Clinic (2020). *Obesidad: Diagnóstico y tratamiento.* https://www.mayoclinic.org/es-es/diseases-conditions/obesity/diagnosis-treatment/drc-20375749 (Accessed 26 March 2020).

McRorie, J. W. (2019). The physics of fiber in the gastrointestinal tract. *Dietary Interventions in Gastrointestinal Diseases* 19–32. doi: 10.1016/B978-0-12-814468-8.00002-8.

Miguel Soca P. E., Sarmiento Teruel Y., Mariño Sole A. L., Llorente Columbié Y., Rodríguez Graña T., Peña González M. (2017). Prevalencia de enfermedades crónicas no transmisibles y factores de riesgo en adultos mayores de Holguín. *Revista Finlay.* 7 (3). [Prevalence of chronic non-communicable diseases and risk factors in older adults in Holguín. Finlay Magazine]

Mishra A. K., Dubey V., Ghosh A. R. (2016). Obesity: an overview of possible role(s) of gut hormones, lipid sensing and gut microbiota. Metabolism 65(1): 48–65. doi: Mishra A. K., Dubey V., Ghosh A. R. (2016). Obesity: an overview of possible role(s) of gut hormones, lipid sensing and gut microbiota. *Metabolism* 65(1): 48–65.

Mohanty, D., Misra, S., Mohapatra, S., Sahu, P. S. (2018). Prebiotics and symbiotic: Recent concepts in nutrition. *Food Bioscience* 26: 152–160. doi: 10.1016/j.fbio.2018.10.008.

Mokhtari, Z., Hekmatdoost, A. (2019). Dietary Interventions and Inflammatory Bowel Disease. *Dietary Interventions in Gastrointestinal Diseases* 33–42. doi: 10.1016/B978-0-12-814468-8.00003-X.

Mugambi, M. N., Young, T., Blaauw, R. (2014). Application of evidence on probiotics, prebiotics and symbiotic by food industry: a descriptive study. *BMC Research Notes* 7(1): 754. doi: 10.1186/1756-0500-7-754.

Nabavi S., Rafraf M., Somi M. H., Homayouni-Rad A., Asghari-Jafarabadi M. (2014). Effects of probiotic yogurt consumption on metabolic factors in individuals with nonalcoholic fatty liver disease. *J Dairy Sci.* 97(12): 7386–93. doi: 10.3168/jds.2014-8500.

Neffe-Skocińska, K., Rzepkowska, A., Szydłowska, A., Kołożyn-Krajewska, D. (2018). Trends and possibilities of the use of probiotics in food production. *Alternative and Replacement Foods* 65–94. doi: 10.1016/B978-0-12-811446-9.00003-4.

Nesterova, A. P., Klimov, E. A., Zharkova, M., Sozin, S., Sobolev, V., Ivanikova, N. V., … Yuryev, A. (2020). Diseases of the digestive system. *Disease Pathways* 443–491. doi: 10.1016/B978-0-12-817086-1.00010-5.

Nguyen, H. T., Truong, D. H., Kouhoundé, S., Ly, S., Razafindralambo, H., Delvigne, F. (2016). Biochemical engineering approaches for increasing viability and functionality of probiotic bacteria. *International Journal of Molecular Sciences* 17(6): 867. doi: doi.org/10.3390/ijms17060867.

Nwaru B., Hickstein L., Panesar S. S., Roberts G., Muraro A., Sheikh A. (2014). EAACI Food allergy and anaphylaxis guidelines group. Prevalence of common food allergies in Europe: a systematic review and meta-analysis. *Allergy* 69(8): 992-1007.

Ojetti V., Ianiro G., Tortora A., D'Angelo G., Di Rienzo T. A., Bibbò S., et al. (2014). The effect of *Lactobacillus reuteri* supplementation in adults with chronic functional constipation: a randomized, double-blind, placebo-controlled trial. *J Gastrointest Liver Dis* 23(4): 387–91. doi: 10.15403/jgld.2014.1121.234.elr.

Olveira G., González Molero I. (2016). Actualización de probióticos, prebióticos y simbióticos en nutrición clínica. *Endocrinología y Nutrición.* 63(9): 482-494. doi: 10.1016/j.endonu.2016.07.006. [Update on probiotics, prebiotics and symbiotics in clinical nutrition. *Endocrinology and Nutrition*]

Park H. J., Lee S. W., Hong S. (2018). Regulation of allergic immune responses by microbial metabolites. *Immune Netw* 18(1): e15. doi: 10.4110/in.2018.18.e15.

Polakowski, C. B., Kato, M., Preti, V. B., Schieferdecker, M. E. M., Campos, A. C. L. (2018). Impact of the preoperative use of symbiotic in colorectal cancer patients: a prospective, randomized, double-blind, placebo-controlled study. *Nutrition* 58: 40-46. doi: 10.1016/j.nut.2018.06.004.

Prince A. C., Myers C. E., Joyce T., Irving P., Lomer M., Whelan K. (2016). Fermentable carbohydrate restriction (low FODMAP diet) in clinical practice improves functional gastrointestinal symptoms in patients with inflammatory bowel disease. *Inflamm Bowel Dis;* 22(5): 1129–36. doi: 10.1097/mib.0000000000000708.

Putta, S., Yarla, N. S., Lakkappa, D. B., Imandi, S. B., Malla, R. R., Chaitanya, A. K., … Aliev, G. (2018). Probiotics: Supplements, food, pharmaceutical industry. *Therapeutic, Probiotic, and Unconventional Foods* 15–25. doi: 10.1016/B978-0-12-814625-5.00002-9.

Rajam, R., Anandharamakrishnan, C. (2015). Microencapsulation of *Lactobacillus plantarum* (MTCC 5422) with fructooligosaccharide as wall material by spray drying. *LWT - Food Science and Technology* 60(2): 773–780. doi: 10.1016/j.lwt.2014.09.062.

Reig-Otero Y., Mañes J., Manyes iFont L. (2017). Sensibilidad al gluten no celiaca (SGNC): manejo nutricional de la enfermedad. *Nutrición Clínica y Dietética Hospitalaria* 37(1): 171-182. doi: 10.12873/371manyesfont. [Non-celiac gluten sensitivity (NCGS): nutritional management of the disease. *Clinical Nutrition and Hospital Dietetics*]

Rogha M., Esfahani M. Z., Zargarzadeh A. H. (2014). The efficacy of a synbiotic containing Bacillus coagulans in treatment of irritable bowel syndrome: a randomized placebo-controlled trial. *Gastroenterol Hepatol Bed Bench* 7(3): 156–63.

Rosado J. L. (2016) Intolerancia a la lactosa. *Gaceta Médica de México.* 152 Suppl 1: 67-73. [Lactose intolerance. *Medical Gazette of Mexico*]

Ruiz Sánchez J. G., Milla S. P., Pelegrina Cortés B., López Plaza B., Bermejo López L. M., Gómez Candela C. (2018). Una visión global de las reacciones adversas a alimentos: Alergia e intolerancia alimentaria. *Nutrición Hospitalaria. Nutr Hosp* 35(4): 102-108. doi: 10.20960/nh.2134. [An overview of adverse reactions to food: Allergy and food intolerance. *Hospital Nutrition. Nutr Hosp*]

Salminen, S., Salminen, E. (1997). Lactulose, lactic acid bacteria, intestinal microecology and mucosal protection. *Scandinavian Journal of Gastroenterology* 32(sup222): 45–48. doi: 10.1080/00365521.1997. 11720717.

Sánchez Salguero E. S., Santos Argumedo L. (2018). La asociación de la microbiota humana con la inmunoglobulina A y su participación en la respuesta inmunológica. Revista Alergia México. *Rev Alerg Mex* 65(3): 264-278. doi: 10.29262/ram.v65i3.519. [The association of the human microbiota with immunoglobulin A and its participation in the immune response. Allergy Magazine Mexico. *Rev Alerg Mex*]

Schumann C. (2002). Medical, nutritional and technological properties of lactulose. An update. *Eur J Nutr.* 41 Suppl 1: 17-25. doi: 10.1007/s00394-002-1103-6.

Scott, K. P., Grimaldi, R., Cunningham, M., Sarbini, S. R., Wijeyesekera, A., Tang, M. L., … Gibson, G. R. (2019). Developments in understanding and applying prebiotics in research and practice – an ISAPP conference paper. *Journal of Applied Microbiology* 128(4): 934-949. doi: 10.1111/jam.14424.

Secretaría de Salud (2016). *Entre 16 y 30 por ciento de la población mexicana padece colitis*. Gobierno de México. https://www.gob.mx/salud/prensa/entre-16-y-30-por-ciento-de-la-poblacion-mexicana-padece-colitis (Accessed 25 March 2020). [*Between 16 and 30 percent of the Mexican population suffers from colitis*. Government of Mexico.]

Secretaría de Salud (2018). *Encuesta Nacional de Salud y Nutrición (ENSANUT)*. Instituto Nacional de Salud Pública. [*National Health and Nutrition Survey (ENSANUT).*]

Secretaría de Salud (2019). 091. *Algunos alimentos provocan hasta el 90% de las alergias alimentarias*. Gobierno de México. https://www.gob.mx/salud/prensa/091-algunos-alimentos-provocan-hasta-el-90-de-las-alergias-alimentarias (Accessed 28 March 2020). [*Some foods cause up to 90% of food allergies*]

Segura-Badilla, O., Lazcano-Hernández, M., Kammar-García, A., Vera-López, O., Aguilar-Alonso, P., Ramírez-Calixto, J., Navarro-Cruz, A. R. (2020). Use of coconut water (*Cocus nucifera* L) for the development of a symbiotic functional drink. *Heliyon* 6(3): e03653. doi: 10.1016/j.heliyon.2020.e03653.

Sengupta S., Koley H., Dutta S., Bhowal J. (2019). Hepatoprotective effects of synbiotic soy yogurt on mice fed high-cholesterol diet. *Nutrition* 63-64: 36-44. doi: 10.1016/j.nut.2019.01.009.

Shah, A., Holtmann, G. (2019). Clinical conditions associated with bacterial overgrowth. *Gastrointestinal Diseases and Their Associated Infections* 67–83.

Sharma, A., Bhagatwala, J. (2019). Food, nutrients, and dietary supplements in management of disorders of gut–brain interaction, formerly functional gastrointestinal disorders. *Dietary Interventions in Gastrointestinal Diseases* 73–80. doi: 10.1016/B978-0-12-814468-8.00006-5.

Shen, J., Zuo, Z. X., Mao, A. P. (2014). Effect of probiotics on inducing remission and maintaining therapy in ulcerative colitis, Crohn's disease, and Pouchitis. *Inflammatory Bowel Diseases* 20(1): 21–35. doi: 10.1097/01.MIB.0000437495.30052.be.

Singh, R. K., Chang, H. W., Yan, D., Lee, K. M., Ucmak, D., Wong, K., … Liao, W. (2017). Influence of diet on the gut microbiome and implications for human health. *Journal of Translational Medicine* 15(1). doi: 10.1186/s12967-017-1175-y.

Soukoulis, C., Yonekura, L., Gan, H. H., Behboudi-Jobbehdar, S., Parmenter, C., Fisk, I. (2014). Probiotic edible films as a new strategy for developing functional bakery products: The case of pan bread. *Food Hydrocolloids* 39: 231–242. doi: 10.1016/j.foodhyd.2014.01.023.

Sperry, M. F., Silva, H. L. A., Balthazar, C. F., Esmerino, E. A., Verruck, S., Prudencio, E. S., … Cruz, A. G. (2018). Probiotic minas frescal cheese added with L. casei 01: Physicochemical and bioactivity characterization and effects on hematological/biochemical parameters of hypertensive overweighed women – A randomized double-blind pilot trial. *Journal of Functional Foods* 45: 435–443.

Staliano, C. D., Martinez, R. C. R., Saad, S. M. I. (2015). Beneficial microorganisms' viability and sensory acceptance of a potentially synbiotic dairy-based tomato spread. *LWT - Food Science and Technology* 62(1): 682–688. doi: 10.1016/j.lwt.2014.12.030.

Sun, L., Pang, Y., Wang, X., Wu, Q., Liu, H., Liu, B., … Jiang, C. (2019). Ablation of gut microbiota alleviates obesity-induced hepatic steatosis and glucose intolerance by modulating bile acid metabolism in hamsters. *Acta Pharmaceutica Sinica B* 9(4): 702-710. doi: 10.1016/j.apsb.2019.02.004.

Taylor, J. (2019). Looking Beyond Lifestyle: A comprehensive approach to the treatment of obesity in the primary care setting. *The Journal for Nurse Practitioners* 9(21): 1555-4155. doi: 10.1016/j.nurpra.2019.09.021.

Tolhurst G., Heffron H., Lam Y., et al. (2012). Short-chain fatty acids stimulate glucagon-like peptide-1 secretion via the G-protein–coupled receptor FFAR2. *Diabetes* 61(2): 364–71. doi: 10.2337/db11-1019.

Tripathi, M. K., Giri, S. K., 2014. Probiotic functional foods: survival of probiotics during processing and storage. *J. Funct. Foods* 9: 225–241. doi: 10.1016/j.jff.2014.04.030.

Ukleja, A. (2018). Nutritional management of gastrointestinal diseases. *Gastroenterology Clinics of North America,* 47(1). doi: 10.1016/j.gtc.2017.12.001.

Uranga J. A., López-Miranda V., Lombo F., Abalo R. (2016). Food, nutrients and nutraceuticals affecting the course of inflammatory bowel disease. *Pharmacol Rep* 68(4): 816–26. doi: 10.1016/j.pharep.2016.05.002.

Vieira da Silva, B., Barreira, J. C. M., Oliveira, M. B. P. P. (2016). Natural phytochemicals and probiotics as bioactive ingredients for functional foods: Extraction, biochemistry and protected-delivery technologies. *Trends in Food Science & Technology* 50: 144–158. doi: 10.1016/j.tifs.2015.12.007.

Waitzberg D. L., Logullo L. C., Bittencourt A. F., Torrinhas R. S., Shiroma G. M., Paulino N. P., et al. (2013). Effect of symbiotic in constipated adult women - a randomized, double-blind, placebo-controlled study of clinical response. *Clin Nutr Edinb Scotl* 32(1): 27–33. doi: 10.1016/j.clnu.2012.08.010.

Waterson M. J., Horvath T. L. (2015). Neuronal regulation of energy homeostasis: beyond the hypothalamus and feeding. *Cell Metab* 22(6): 962–70. doi: 10.1016/j.cmet.2015.09.026.

World Health Organization (WHO) (2015). *Obesity.* http://www.who.int/topics/obesity/en/(Accessed 26 March 2020).

Yamamoto-Furusho, J. K., Bosques-Padilla, F. J., Charúa-Guindic, L., Cortés-Espinosa, T., Miranda-Cordero, R. M., Saez, A., Ledesma-Osorio, Y. (2020). Epidemiología, carga de la enfermedad y tendencias de tratamiento de la enfermedad inflamatoria intestinal en México. *Revista de Gastroenterología de México.* RGMX-616:11. doi: 10.1016/j.rgmx.2019.07.008. [Epidemiology, burden of disease, and treatment trends for inflammatory bowel disease in Mexico. *Journal of Gastroenterology of Mexico.*]

Yamamoto-Furusho, J. K., Bosques-Padilla, F., de-Paula, J., Galiano, M. T., Ibañez, P., Juliao, F., … Zaltman, C. (2017). Diagnóstico y tratamiento de la enfermedad inflamatoria intestinal: Primer Consenso Latinoamericano de la Pan American Crohn's and Colitis Organización. *Revista de Gastroenterología de México* 82(1): 46–84. doi: 10.1016/j.rgmx.2016.07.003. [Diagnosis and treatment of inflammatory bowel disease: First Latin American Consensus of the Pan American Crohn's and Colitis Organization. *Journal of Gastroenterology of Mexico*]

Yamamoto-Furusho, J. K., Gutiérrez-Grobe, Y., López-Gómez, J. G., Bosques-Padilla, F., Rocha-Ramírez, J. L. (2018). Consenso mexicano para el diagnóstico y tratamiento de la colitis ulcerosa crónica idiopática. *Revista de Gastroenterología de México* 83(2): 144-167. doi: 10.1016/j.rgmx.2017.08.006. [Mexican consensus for the diagnosis and treatment of idiopathic chronic ulcerative colitis. *Journal of Gastroenterology of Mexico*]

Zheng, X. E., Mullin, G. E., Salem, A. E., Tomakin, E., Roland, B. C. (2015). Obesity increases risk of small intestinal bacterial overgrowth (SIBO). *Gastroenterology* 148(4): S–673. doi: 10.1111/nmo.13199.

Chapter 2

Polyphenols Nanoformulations for Cancer Therapy and Apoptosis-Inducing Molecular Mechanisms

Beatriz del C. Coutiño-Laguna, Adriana C. Flores-Gallegos, Juan A. Ascacio-Valdés, Anna Iliná, Aidé Sáenz-Galindo, Adali O. Castañeda-Facio, Sandra C. Esparza-González and Raúl Rodríguez-Herrera*

Department of Food Research, Autonomous University of Coahuila, Coahuila, México

Abstract

Cancer is defined as a group of multifactorial diseases caused by environmental and genetic factors that induce a series of mutations in normal cells that cause abnormal cell growth with possible malignant and fatal neoplasties. Polyphenols are natural phytochemicals that have been widely used for the prevention and treatment of various ailments due to their anti-glycemic, anti-inflammatory, antioxidant, anticancer, and cytotoxic effects; thus, they could potentially be capable of acting as chemotherapeutic and chemopreventive agents in the treatment of cancer. However, these phytochemicals have low bioavailability.

* Corresponding Author's E-mail: raul.rodriguez@uadec.edu.mx.

In: Food Processing
Editors: Myriam Huijs
ISBN: 978-1-68507-570-5

Therefore, nanoparticles (NPs) are likely to provide an attractive new alternative to conjugate the efficacy of various bioactive phytochemicals in cancer therapy. This is due to the physicochemical, optical, magnetic, electrical, mechanical properties, surface area, size, and dispersion that these nanostructures have, which would provide an innovative green nanotechnology approach to the use of polyphenols to obtain nanoparticles. In this review, we have analyzed the most current data on the different polyphenol nanoformulations as alternative chemotherapeutic and chemopreventive treatments in the treatment of cancer and discuss their multiple actions and anti-cancer molecular mechanisms that induce apoptotic potential in cancer cells.

Keywords: apoptosis, cancer, molecular mechanisms, nanoparticles (NPs), polyphenols

Introduction

Cancer is a group of multifactorial diseases caused by environmental and genetic factors that induce a series of mutations in normal cells, losing control of cell division, resulting in uncontrolled cell proliferation (Albulet et al., 2017). However, for this process, energy and biomass production are needed to achieve cell maintenance and expansion (Reyes-Farias & Carrasco-Pozo, 2019). Cancer cells spread by metastasis to other parts of the body to invade healthy tissues which put at risk to the healthy of the patient. Therefore, cancer is the principal risk factor of mortality and morbidity of people. The most frequent cancers are cancer breast, prostate, lung, liver, and colorectal cancer (Buttacavoli et al., 2018; Ovais et al., 2016). Common treatments to combat cancer include radiation therapy, cryotherapy, hyperthermia, stem cells, and surgery as non-pharmacological treatments but these are unspecific and unsatisfactory, pharmacological treatments include immunotherapy, hormonal laser therapy, and chemotherapy (Davatgaran-Taghipour et al., 2017). However, these treatments have limitations as they are costly, generate drug resistance (Riahi-Chebbi et al., 2019), and are not selective in attacking cancer cells, such as chemotherapy that affects normal cells causing multiple toxic side effects such as neuropathy, tiredness, abnormal immune system function, and hair loss, with few reductions of metastasis, little improvement and deteriorating quality of life of patients (Alsalhi et al., 2016; Alshatwi, A., et al., 2015b, 2015a; Badea et al., 2020).

Nanotechnology is a useful tool to treat cancer, this discipline is focused on designing, controlling, and functionalizing matter at the nanoscale within the range of ≤ 100 nm, which do it suitable for the treatment of treatem cancer (Chugh et al., 2018; Fariq et al., 2017; Gómez-Garzón, 2018). Through nanotechnology, nanostructures with high surface area ratio (volume/surface ratio) and photocatalysis properties can be obtained (Davatgaran-Taghipour et al., 2017; Friehs et al., 2016). These properties allow those nanostructures to change their physicochemical properties such as their shaper and, stability, allowing them to be compatible with matrices biological (Abdullaeva, 2017). Nanotechnology has shown promising effects in the treatment of cancer with few adverse effects through the uses of NPs for targeted drug delivery to the target site (neoplasms) (Davatgaran-Taghipour et al., 2017). NPs can be a vector for the administration of drugs or polyphenols that ensure the delivery of the compounds to the target site, and this allows to be controlling the expression and function of apoptotic processes (Albulet et al., 2017; Alshatwi, A., et al., 2015b).The nanomaterials main used in biomedicine are polymeric, lipid, carbon-based NPs, inorganic NPs, and metallic NPs (Son et al., 2016).

Polyphenols are secondary metabolites of plant origin with the capacity to exert health-promoting activities that include anti-inflammatory, antioxidant, and antineoplastic activities (Davatgaran-Taghipour et al., 2017; Williamson, 2017). *In vivo* and *in vitro* studies have shown that polyphenols inhibit the development of different types of cancer such as liver, breast, intestine, colon, prostate, and bladder, among others (Alshatwi, A., et al., 2015b; Liang et al., 2011). The polyphenols that stand out for having the bioactivity of inhibiting carcinogenesis are carotenoids, flavonoids, curcumin, quercetin, resveratrol, and phenolic acids, epigallocatechin 3-gallate, baicalein, chrysin, silibinin, honokiol, and luteolin (Alshatwi, A., et al., 2015b; Davatgaran-Taghipour et al., 2017; Williamson, 2017). Functionalization of the NPs with polyphenols has shown synergistic effects since these in conjunction have shown inducing apoptosis (Alsalhi et al., 2016; Alshatwi, A., et al., 2015b; Davatgaran-Taghipour et al., 2017). One way to obtain these biohybrid materials is through green synthesis which is an economical, efficient, and ecological method (Devanesan et al., 2018; Noah, 2019). The aim of the current study is to review the currently available data on the on the different nanoparticles metal and metal oxide functionalized through the green synthesis with extracts of plant-based and the molecular mechanisms through which apoptosis is induced on tumor cells.

Cancer: A Global Health Hazard

Cancer refers to a group of multifactorial, non-communicable diseases caused by environmental and genetic factors, which trigger a series of mutations in normal cells that result in abnormal growth of malignant cells with the potential to spread to invade or reach other parts of the body (metastasis), potentially leading to the death of the patient. (Buttacavoli et al., 2018; Montané et al., 2020; Sánchez, 2013). This phenomenon starts during the cell cycle in the mitosis phase and usually takes place in a controlled manner in the G0 phases and at or at checkpoints in the G1/S or G2/M phases. However, some cells begin to divide uncontrollably. This event may be origin by overexpression of oncogenes in a metabolic environment optimal for their development (oxygen, glucose). Thus, after a healthy cell gives way to the initiation, promotion, and progression it becomes a cancer cell, establishing a targeted tumor and subsequently causing metastasis (Vervandier-Fasseur & Latruffe, 2019).

This disease is extremely devastating around the world, being considered one of the major public health issues with a significant global impact in both developed and developing countries, as well as being the most important barrier to increasing life expectancy in all countries of the world (Bray et al., 2018; Choudhari et al., 2020). In 2018, there were an estimated 9.6 million cancer deaths and 18.1 million new cases of cancer, and according to the American Cancer Society (ACS), it is estimated that by 2030 there will be 23.6 million new cases of cancer, and by 2040 an increase of 27.5 million new cases of cancer is expected. The types of cancer that cause the highest mortality are breast, prostate, lung, liver, and colorectal cancer. However, the types of cancer that cause the highest mortality and incidence tend to vary between countries (Bray et al., 2018; Choudhari et al., 2020; Ovais et al., 2016). In cancer therapy, the traditional treatments are surgery, radiotherapy, immunotherapy, and among the most widely used is chemotherapy as a single or successive treatment (Vervandier-Fasseur & Latruffe, 2019). However, the main disadvantages of chemotherapy are the use of high doses that cause drug resistance, leading to cancer recurrence (Choudhari et al., 2020). Moreover, it does not attack target sites, as it not only acts on cancer cells but also healthy cells and therefore affects the patient's quality of life, causing adverse effects that deteriorate the patient's quality of life (Badea et al., 2020; Thipe et al., 2019). To overcome the problems of current cancer therapies the search continues for promising new anticancer agents with better efficacy and fewer side effects (Choudhari et al., 2020).

Natural Polyphenols: Sources for Cancer Therapy

The progressive understanding of some molecular mechanisms involved in the development of cancer has led to the development of various anticancer drugs, intending to decrease the mortality rate by preventing or treating cancer in its early stages. However, chemically synthesized drugs have not provided significant improvements in survival rates in recent years. Efforts have therefore focused on the search for new strategies and chemopreventive agents that overcome the problems of current therapies or serve as a complement to improve their efficiency and with fewer side effects (Choudhari et al., 2020; Martínez et al., 2021).

Natural plant compounds known as polyphenols (phytochemicals) serve as vital resources for new drugs and are also sources for cancer therapy (Choudhari et al., 2020). Polyphenols are produced by the secondary metabolism of plants (Guía-García et al., 2021) through different routes such as shikimic acid and acetic acid (Mercado-Mercado et al., 2013; Rojas, 2012), as part of their defense mechanism to act as sweeps extreme weather conditions and pathogens (Guía-García et al., 2021). These compounds have a wide range of complex structures that are classified according to their nature. The basic monomer in polyphenols is the phenolic ring, and depending on the strength of the phenolic ring, polyphenols can be classified into flavonoids, phenolic acids, phenolic alcohols, stiblins, and lignans (Figure 1) (Abbas et al., 2017; Papuc et al., 2017). These compounds are implicated in exerting biological activities with protective effects on human health, as they can interact with enzyme systems involved in crucial pathways of living organisms exerting antiviral, antibacterial, cardioprotective, anti-inflammatory, antidiabetic, anticancer, and antiallergic activities (Abbas et al., 2017; Ozturk et al., 2018; Peñarrieta et al., 2014; Wang et al., 2018). In this sense, the biological activities of polyphenols could be correlated to their chemical structure, their aromatic rings, and in particular to the OH groups on the aromatic (phenyl) ring (Castro-Acosta, 2019; Kabera et al., 2014; Selim et al., 2020; Wang et al., 2018; Yoosefian et al., 2019).

Because of all the health benefits of polyphenols, in recent years there has been increasing interest in incorporating plant-based foods as they naturally contain polyphenols which act as natural antioxidant products, as well as providing anticarcinogenic activities (Abbas et al., 2017). However, polyphenols are present in plant matrices in free and bound forms (as a structural component of the cell wall and covalently bound to the sugar parts) (Tang et al., 2021), can be extracted by simple, green techniques such as ultrasound-assisted extraction, microwave-assisted extraction, ohmic heating (OH), and supercritical fluid (Guía-García et al., 2021). Through these extraction techniques, the polyphenols retain most of their properties intact and could contribute to the study of these compounds as promising alternative drugs against cancer (Montané et al., 2020).

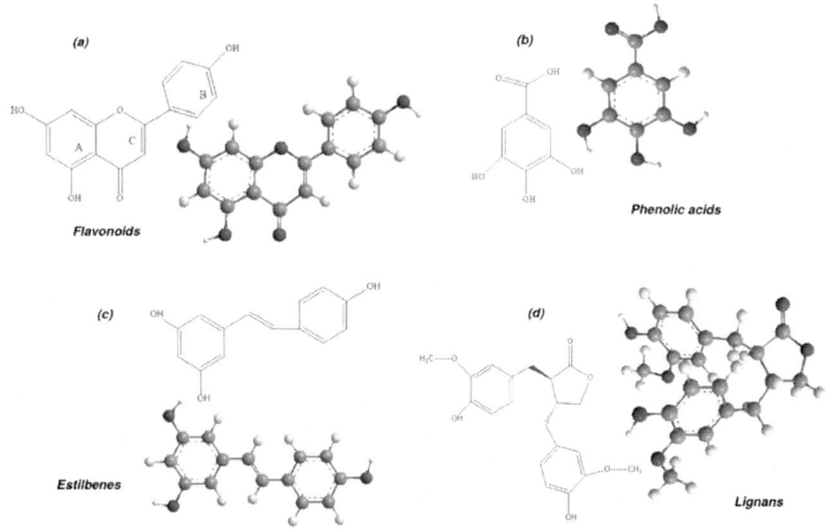

Figure 1. Chemical structures of the main groups of polyphenols, as the first main group is a) flavonoids, followed by b) phenolic acids, c) stilbenes and d) lignans.

Polyphenols may act as immunomodulatory molecules (the ability to modulate immune responses by inducing, amplifying, attenuating, or preventing the spread of cancer cells). Numerous studies have shown that polyphenols have direct antitumor properties by modulating various signaling pathways that are out of sync in cancer cells (Trung & An, 2018). Cellular mechanisms include activation of caspase enzymes that induce cancer cell cycle arrest, reducing tumour vasculature, reducing tumour cell invasion and metastasis, and induction of mitochondrial damage, and finally apoptosis of the neoplasm (Davatgaran-Taghipour et al., 2017). An example of this is the quercetin that demonstrated inhibition of lung, colon, breast, prostate, ovarian, and endometrial cancer cells (Daneshmehr, 2015). Thus, quercetin when absorbed by cancer cells can exert antioxidant activity, which is reflected in the enzymatic activity present, the activation of glutathione, the detection of reactive oxygen species (ROS), and the regulation of signal transduction pathways, such as NRFB, MAPK, and AMPK, which arrest the cell cycle, promote apoptosis, inhibit angiogenesis and metastasis *in vitro* (Tang et al., 2020). In addition, quercetin through molecular modulation measured by PI3K, Akt, mTOR, Wnt, β-catenin, and MAPK pathways, ERK1/2, promotes autophagy, loss of cell viability, and apoptosis during cancer metabolism, specifically in aerobic glycolysis metabolism and mitochondrial function, due to the antioxidant activity it exerts with chemopreventive effects (Reyes-Farias & Carrasco-Pozo, 2019). So too, resveratrol (RSV) showed antiproliferative effects in MSTO-211H (malignant mesothelioma) (Lee et al., 2019). On the other hand, epigallocatechin-3-gallate (EGCG) (Mukherjee et al., 2015), resveratrol (RSV), and fisetin (FS) induce a dose-dependent reduction on SH-SY5Y-CFP-DEVD-YFP cell viability that is likely to involve the activation of the apoptotic pathways (Sanna et al., 2014). In the same way, reports indicate that the flavonol Kaemferol has demonstrated the ability to inhibit cell growth in human breast cancer (MDA-MB-453) through the havoc of the cell cycle in the G2/M stage through the downregulation of CDK1 (cyclin-dependent kinase 1) in breast cancer cells. In addition, Kaemferol destabilizes the uptake of glucose in the cancer cell, which causes the capacity for cell proliferation and reduced cell viability (Wang et al., 2019). Likewise, this flavonol can cause a break in the PARP protein (Poly ADP ribose polymerase) which repairs damage at the DNA level. There is also the expression of Bcl2 and Bax, which indicate apoptosis in breast cancer cells (MCF-7) according to what was mentioned by Yi et al., (2016). Similarly, the coumarins, curcumins, lignans, flavonolignans, flavonoids, naphthoquinones, and stilbenes have shown anticancer activities (Davatgaran-Taghipour et al., 2017).

Nanoparticles in Cancer Treatment

The importance of research into cancer-related diseases has also been of interest to nanomedicine, which allows the application of nanopharmaceuticals treatments through nanotechnology (Ali et al., 2020). The use of nanotechnology offers various clinically useful tools and devices that allow improving efficiency and addressing some limitations of drug administration in cancer (Montané et al., 2020). Consequently, nanotechnology develops structures and systems in the $\leq$100 nm range to control their shape and size at nanometre scales (Abdullaeva, 2017; Albulet et al., 2017), with the potential to be applied at molecular levels as there is a good biological interaction between nanoscale materials and biological macromolecules such as vitamins, enzymes, hormones, DNA plasmids, and antibodies, due to the similar size of nanostructures and molecules (Albulet et al., 2017; Montané et al., 2020; Ovais et al., 2016).

Applications of nanotechnology in the field of biomedicine involve the use of nanoparticles (NPs) with sizes from 1 to 100 nm (Mirza & Siddiqui, 2014), including the use of polymeric nanoparticles (PNPs), lipid NPs, carbon-based NPs, inorganic NPs such as quantum dots (QDs), metal oxide NPs (MONPs) such as SiO_2, ZnO, TiO_2, CeO_2, $Al2O_3$, or Fe_2O_3 NPs, and metallic NPs such as AuNP, AgNP, CuNP, FeNP, ZnNP, PtNP, and PdNP, which are used in cancer therapy and diagnostics (De Matteis, 2017; Khan et al., 2019; Wang et al., 2020).

Among all the nanoparticles mentioned, inorganic nanoparticles have gained great interest as they show electrical properties, good stability, biocompatibility, low toxicity. (Montané et al., 2020), thermal capacity, modification of its surface, chemical reactivity, oxide-reduction (Abdullaeva, 2017), magnetic, mechanical, and catalytic properties (Gurunathan et al., 2015; Iqbal et al., 2018), with photocatalysis properties, greater surface area (high volume/surface ratio) (Friehs et al., 2016), and ability to modify its surface (Abdullaeva, 2017), optical properties (López-Gomez., 2017), surface plasmon resonance (SPR) (Bethu et al., 2018), hsuggest that nanoparticles are ideal candidates for use in medical activities, since some drugs to treat cancer can be functionalized to the surface of NPs through ionic or covalent bonds (Montané et al., 2020). In addition, NPs may be promising vectors for the transport and delivery of drugs for cancer diagnosis and therapy. (Davatgaran-Taghipour et al., 2017; Mirza & Siddiqui, 2014), as they induce targeted delivery, increase the stability of drugs, and protect them from degradation within the body during their passage through biological fluids, to enable them to be pharmacologically active compounds and to reach the target site at optimal rates in adequate concentrations (Albulet et al., 2017; Fariq et al., 2017).

Nanoparticles Functionalized with Polyphenols to Treat Cancer

Silver nanoparticles (AgNPs) have excellent properties such as chemical stability in aqueous solutions, antimicrobial activity, catalytic and fungicidal activity, small size, high thermal conductivity, biocompatibility, and potential ability to scavenge free radicals (Rasheed et al., 2017; Zhu et al., 2016).

Therefore, AgNPs have excellent characteristics for the treatment of cancer since some studies show how AgNPs can enter cancer cells and inhibiting their viability by generating reactive oxygen species (ROS) (He et al., 2018) by altering the respiratory chain mitochondrial. The AgNPs also cause apoptosis, inhibiting cell migration and invasion (Buttacavoli et al., 2018; Yuan et al., 2017), as well as the induction of lactate dehydrogenase (LDH) leakage (Gurunathan et al., 2018), damage at the DNA level (Dehghanizade et al., 2018; Govender et al., 2013; Singh et al., 2017). Finally, they induce the inhibition of the proliferation of tumor cells (Albulet et al., 2017). In this context, AgNPs play an important role as an anticancer agent. However, the effects will depend on the shape of the AgNPs, which can be spherical, hexagonal, or irregular, the sizes, types of coatings, and doses administered (Wei et al., 2015). In addition, AgNPs exert great antimicrobial potential in the treatments of wounds, burns, or bacterial infections (Singh et al., 2017).

On the other hand, gold nanoparticles (AuNPs) also have several advantages, including nanometric sizes (between 1 and 150 nm), exceptional electrical, optical, chemical, sensing, and biochemical properties. In addition, they can modify their shape, size, surface area, are easy to synthesise, stable, and have high biocompatibility along with low toxicity, and surface reactivity, making AuNPs with biological media possible. Therefore, AuNPs are potentially effective in drug delivery and tumour therapy in the early detection of cancer (Albulet et al., 2017; Montané et al., 2020).

The use of AuNPs is considered safe for biological treatments according to the approval of the FDA (US Food and Drug Administration), since AuNPs have an inert and non-toxic core, for this reason, they can be used as carriers in the administration of drugs (Raghunandan et al., 2011; Yuan et al., 2017). They allow prolonging the activity of antineoplastic drugs given their structure spherical, wide surface, and biocompatibility (Zhao, Chengbo, et al., 2018; Zhao, Xia-Mei, et al., 2018). AuNPs have a better match with the neo-vasculature of certain tumors (Shukla et al., 2012), causing lethal damage to the cancer cell membrane surface due to the influence of free e- (electrons) (Raghunandan et al., 2011). Likewise, they are promoters of apoptosis in tumor cells through the peroxidation of lipids that alter the normal cell cycle, and the breakdown of the mitochondrial membrane causing DNA fragmentation and regulating the expression of some genes such as hypoxia inducible- factors (HIFs) and the proto-oncogene c-Myc (Bodelón et al., 2017).

Also, the AuNPs, exert modulatory effects on the inflammation that plays an important role in various diseases such as cancer, since inflammation is a complex biological process of the body´s response to different aggressions, through decreased transiently the pro-inflammatory cytokine levels (IL-1α, IL-1β, MCP-1) and increased the anti-inflammatory IL-2 secretion, in accordance with the study of Filip et al., (2019).

Green Synthesis of Nanoparticles

The surface functionalization of inorganic nanoparticles gives them good compatibility and versatility, making these nanostructures ideal candidates for use in biomedical applications (Montané et al., 2020). Therefore, inorganic nanoparticles have excellent properties such as chemical stability in aqueous solutions (Albulet et al., 2017). Thus, they have attracted attention for their use in the delivery of polyphenols by functionalization to doped polyphenols to enhance their unique properties, as well as providing a suitable environment to preserve polyphenol activity, enhance biological effects and improve electron transport (Alsalhi et al., 2016; Khan et al., 2019; Raymundo-Pereira et al., 2020). The functionalization process is carried out from the synthesis of nanoparticles by different methods such as physical or chemical, which allow modifying the surface of nanomaterials, reducing metal ions, achieving the addition of ligands to give stability, and obtaining the synthesis of NPs (Singh et al., 2017). However, they are carried out using expensive, toxic solvents and surfactants, which generate hazardous waste that damages the environment. Moreover, they are labor-intensive, costly, time-consuming, and energy-intensive methods (Bethu et al., 2018; Gómez-Garzón, 2018). However, recent research indicates that biological methods play an important role in the synthesis of nanoparticles, where "bionanofactories" (biological agents' bacteria, fungi, viruses, or algae) can be used (Noah, 2019; Ovais et al., 2016). But, green synthesis involving plant matrices using aqueous extracts (AE) (loaded with polyphenols) provides a more suitable approach, as they act as stabilizing and reducing agents in the functionalization of the NPs during green synthesis (Iqbal et al., 2020), thanks to the immobilization and capture of metal ions which are reduced ($M^+ \rightarrow M^\circ$) to biosynthesize and obtain biologically active NPs (Ovais et al., 2016) as illustrated graphically in the Figure 2.

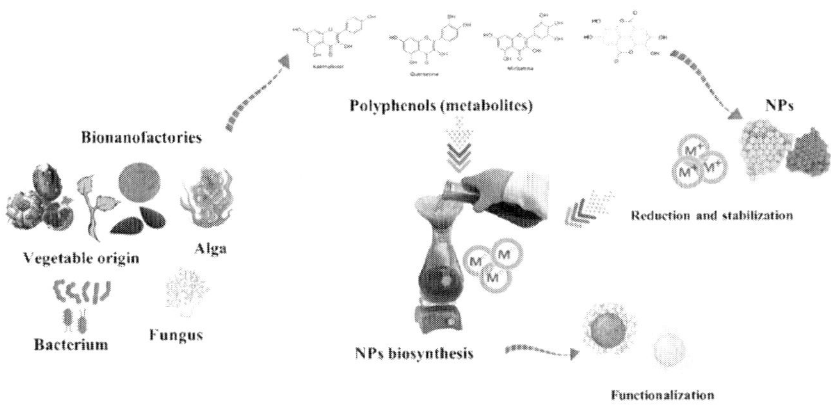

Figure 2. Schematic presentation of the funcionalization of NPs using different bionanofactories (plants, bacteria's, fungi, and algae) through the green synthesis of nanoparticles (NPs).

The Role Ofpolyphenol Nanoformulations and Advances in Cancer

In recent years, nanoparticles have been investigated for cancer therapy and diagnosis, using the phytonanotechnology approach. Several studies have demonstrated the anticancer activity of various plants (rich in polyphenolic compounds), which have been functionalized with inorganic nanoparticles through green synthesis using their aqueous extracts (Ovais et al., 2016) as well as being tested in a variety of cancer cell lines such as MCF-7 (human breast cancer), HEPG2 (human liver cancer), AGS (human prostate cancer), PC-3 (human prostate cancer), HeLa (cervical cancer), HCT-116, HCT-15 (human colon adenocarcinoma), Hep-G2 (liver carcinoma), Caco-2 (intestinal adenocarcinoma), A549 (lung cancer), MBT-2 (murine bladder tumor cells-bladder cancer), SGC-7901 (Human gastric carcinoma) (Alsalhi et al., 2016; Bethu et al., 2018; Dehghanizade et al., 2018; Govender et al., 2013; Hsieh et al., 2012; Lee et al., 2019; Mukherjee et al., 2015; Qian et al., 2019; Raghunandan et al., 2011; Shukla et al., 2012; Yuan et al., 2017; Zhang et al., 2019; Zhu et al., 2016).

Table 1 extensively details the results of gold and silver nanoparticles functionalized with aqueous plant extracts by green synthesis and the results obtained against cancer. These studies have shown that polyphenol nanoformulations exert direct anti-tumor properties by modulating multiple signaling pathways that are deregulated in various cancer cells. In addition, it should be noted that many of these studies used healthy cells to differentiate cytotoxic damage from cancer cells, for example, the normal cell line L929 (healthy mouse fibroblasts) (Dehghanizade et al., 2018; Menon et al., 2012), PBMC (human peripheral blood mononuclear cells) (Seo et al. 2014), HEK-293 (healthy human kidney cells) (Raghunandan et al., 2011), CCD112 (normal colon cells) (Lee et al., 2019), RAW264.7 (murine macrophage) (Oh et al., 2018; Wang et al., 2017), hSSCs (Neonatal skin stromal cells) (Alsalhi et al., 2016), and the HaCaT cell line (immortal spontaneously transformed immortal keratinocytes from adult human skin) (Yuan et al., 2018), to mention a few cases where functionalized NPs did not cause significant cytotoxic damage to normal healthy cells.

NPs in the Human Body and Their Elimination

It is essential to mention that at present there are various concerns about the toxicity of nanoparticles and the secondary effects that they could cause on healthy cells due to their small size, surface area, the cellular transport in which they intervene, the formation of free radicals and the agglomerations of NPs or their derivatives (Albulet et al., 2017; Siddiqui et al., 2017). However, several studies show experimentally (*in vitro* and *in vivo*) that NPs synthesized with polyphenolic compounds only cause damage to cancer cells while protecting healthy cells from adverse effects as mentioned throughout this review. Although, the effects could vary significantly under in vivo conditions (Ovais et al., 2016; Qian et al., 2019; Zhang et al., 2019).

Table 1. List of studies that take advantage of plant aqueous extracts for the green synthesis of NPs as therapeutics agents for to treat cancer

Type of NPs	AE	Nanoformulation	Size ranging (nm)	Shape	IC50 Value	Cancer cell lines	Results	Reference
AgNPs	*Pimpinella anisum* L. (Pim) seeds	Pim-AgNPs	8.3 nm	Spherical	>10 μg/ml to 50 μg/ml	HT115 and hSSCs	Necrosis (10 g/ml), reduced mitochondrial function (15 nm more toxic), membrane damage, ROS production (50 g/ml), oxidative stress.	(Alsalhi et al., 2016)
AgNPs	*Garcinia mangostana* (GM) fruit peels	GM-AgNPs AgPCA (rotocatechuic acid (PCA) was used as a test drug).	32.7 ± 5.7 nm	Spherical	15.6 μg/ml	HCT116	Mitochondrial membranes damage, increase in reactive oxygen species (ROS)	(Lee et al., 2019)
AgNPs	Vitamin C (VC)	VC-AgNPs	2 nm	Monodisperse spherical	5, 10 and 20 μg/ml	HePG-2	Apoptosis, Disrupting MAPK, overproduction of ROS, DNA fragmentation, caspase-3 activation, AKT, MAPKs and p53.	(Zhu et al., 2016)
AgNPs	Extract of *Rhynchosia suaveolens* (RS)	RS-AgNPs	10-30 nm	Spherical	4.35, 7.72, 4,2 and 24.7 μg/ml	A549, SKOV3, DU145 and PC-3	Oxidative stress, increased LPO, ROS, Membrane damage, DNA and cytoplasmic	(Bethu et al., 2018)

Table 1. (Continued)

Type of NPs	AE	Nanoformulation	Size ranging (nm)	Shape	IC50 Value	Cancer cell lines	Results	Reference
							condensation, low in GSH levels. Finally, apoptosis, by the regulation of p53, caspases-3, caspase-8, caspase-9, Bax, Bcl-2, increasing the activity of caspase 3/7.	
AgNPs	*Anthemis atropatana* (Aa) extract	Aa-AgNPs	38.89 nm	Spherical	100 µg/ml	HT29	DNA fragmentation and ROS production, Bax and Bcl-2 gene expression, causing apoptosis.	(Dehghanizade et al., 2018)
AgNPs	*Nelumbo nucifera* (Nn) extract	Nn-AgNPs	10 to 15 nm	Spherical	17.8 µg/mL and 19.8 µg/mL	SGC-7901, PC-3 and HepG2	Production of ROS, apoptosis by interaction with proteins, cellular lipids, and DNA.	(He et al., 2018)
AuNPs	RSV	RSV-GNPs	39 nm	Spherical	8 µg/mL (*in vitro*) and 16.5 mg/kg (*in vivo*)	Hepg2 (*in vitro* and *in vivo* treatment in mice). The Hepg2 cells were prepared (5×10^6 viable cells/mL).	Promoted the apoptosis of Hepg2 cells, because of by down-regulating pro-caspase-3, pro-caspase-9, Akt, PI3K and upregulating bax and caspase-8.	(Zhang et al., 2019)

Type of NPs	**AE**	**Nanoformulation**	**Size ranging (nm)**	**Shape**	**IC50 Value**	**Cancer cell lines**	**Results**	**Reference**
						The mice were injected subcutaneously in the right flank with 0.2 ml of the cell suspension tumoral.	With an inhibition of tumor growth of 57.2% (*in vivo* treatment)	
AuNPs	Fresh tea EGCG, FU-CMC	(PNB) (FU-CMC-EGCG-PNB)	30-70 nm	Spherical	10, 20 and 30 mg/l	HaCaT ((*in vitro*) and MKN-45 (*in vivo* treatment in mice) tumor cells were immunized through the implantation	Gastric anticancer effect in the in vivo study. Cytotoxic and apoptotic effects	(Yuan et al., 2018)
AuNPs	Garlic estracts	G-AuNPs	85±5 nm	Spherical and anisotropic	500 μg/ml	MCF-7 and L929	MCF-7 cytotoxic effects	(Menon et al., 2012)
AuNPs	EGCG and Green tea	EGCG-NPs and GT-NPs	20-50 nm	Spherical	2 μg / ml and 20 μg / ml	MCF-7	Prompted apoptosis by increasing reactive oxygen species (ROS), depleting the antioxidant defense mechanism, decreased Bax/Bcl2 ratio and active Caspase 3 levels, decreased NF-κB	(Mukherjee et al., 2015)

Table 1. (Continued)

Type of NPs	AE	Nanoformulation	Size ranging (nm)	Shape	IC50 Value	Cancer cell lines	Results	Reference
AuNPs	EGCg	EGCg-AuNPs	11 mm	Spherical	136 μCi and 30 μL (IT)	*in vivo study* of radiotherapy in SCID mice with PC-3	In PC-3 mice, it reduced 80% of tumor volumes, with significant inhibition of tumor growth, and affinity for Lam 67R receptors, which are over-expressed in prostate tumor cells	(Shukla et al., 2012)
AuNPs	Guava leaf *(Psidium guajava)* extract and Clove bud *(Syzygium aromaticum)* extract	AuNPs-Guava AuNPs-Clove	25-30 nm and 5-100 nm	Spherical, polyshaped and irregular	10, 20 and 30 μg/ml	HT-29, HEK-293, K-562, and HeLa	Cell proliferation inhibition generation of free radicals and cell damage	(Raghunandan et al., 2011)
AuNPs	*Alternanthera sessilis* aqueous extract	As-AuNPs	12–18 nm	Spherical	1-15 μg/ml	A549 and HeLa	Decreased cell viability, apoptosis by activation of caspase-3, 8 and 9 mediated, increased expression of Bax and decreased expression of Bid and Bcl-2	(Qian et al., 2019)

Type of NPs	**AE**	**Nanoformulation**	**Size ranging (nm)**	**Shape**	**IC50 Value**	**Cancer cell lines**	**Results**	**Reference**
AuNPs	EGCG	EGCG-AuNPs	50 nm	Spherical	2 mg for mice	*In vivo* and *in vivo* study of MBT-2 with C3H/He female mice. MBT-2 tumor cells (5×10^5) in a volume of 50 μL were injected subcutaneously into the right dorsal flank of the mice.	Promoted apoptosis, decreased cell viability, arrest in the S phase of the cell cycle, measured by Bcl-XL, Bad, Bax. Finally, it reduced tumor growth, less accumulation of fluid in the peritoneal cavity	(Hsieh et al., 2012)
AuNPs	RSV, GA	RSV-AuNPs, RSV-GA-AuNPs	<200 nm	Mono-dispersed and spherical	16.181 μg/mL to 18.350 μg/mL	MDAMB-231, PANC-1 and PC-3.	Introduction by endocytosis, apoptosis (nuclear condensation, cell contraction, fragmentation, and dysfunction)	(Thipe et al., 2019)
Au-Nps and AgNps	*Chaenomeles sinensis* (Cs)	Cs-AuNPs and Cs-AgNPs	20–40 nm and 5–20 nm	Spherical	0-50 μg/ml and 0-10 μg/ml	MCF7 and RAW264.7	Reducing cell viability and inhibition of cancer cells and showed no significant cytotoxic damage in RAW264.7 cells	(Oh et al., 2018)

Table 1. (Continued)

Type of NPs	AE	Nanoformulation	Size ranging (nm)	Shape	IC50 Value	Cancer cell lines	Results	Reference
Au-NPs Ag-NPs	Aqueous extract of *Cibotium Barometz* (CB) root	CB-AuNPs and CB-AgNPs	6 nm and 23 nm	Spherical	0.1-10 μg/ml	MCF-7 and RAW264.7	Cytotoxicity of cancer cells (MCF-7) and did not exhibit cell death of RAW264.7 cells	(Wang et al., 2017)

Cancer cell lines: MDAMB-231 (Human breast), PANC-1 (Pancreatic cancer), HT115 (Colon cancer cells), HCT116 (Colorectal cancer), A549 (Lung cancer cells), HePG-2; HepG2 (Human hepatocellular carcinoma cells), MCF-7 (Breast cancer), HT29 (Colon cancer cell), PC-3 (Human prostate carcinoma cell lines), SKOV3 (Human ovarian carcinoma), HeLa (Cervical cancer cell line), SGC-7901 (Human gastric carcinoma), PC-3 (Human prostate cancer) and HepG2 (Human liver cancer). Normal healthy cells: hSSCs (Neonatal skin stromal cells), HaCaT (Spontaneously Transformed Immortal Aneuploid Keratinocytes from Adult Human Skin), L929 (mouse fibroblast cells), HEK-293 (Human kidney), murine macrophage (RAW264.7). Abbreviations: AE (aqueous extract), ROS (Reaction oxygen species), EGCG (Polyphenol: (-) - epigallocatechin-3-gallate), EGCg (Epigallocatechin-gallate), FU-CMC (Fucose-carboxymethyl chitosan), AgNPs (Silver nanoparticles), AuNPs (Gold nanoparticles), Res (Resveratrol), RSV (Resveratrol)·FS (Fisetin), LDH (Lactate dehydrogenase leakage), MDA (Malondialdehyde), NAD (Nicotinamide Adenine Dinucleotide), DNA (Deoxyribonucleic acid), PARD-1 (Poli (ADP-Ribosa) Polimerasa-1), LPO (Lipid peroxidation), MAPK (Mitogen-Activated Protein Kinases), AKT (protein kinase B), PI3K (phosphatidylinositol-3-kinase), AKT (protein kinase B), DOX (Doxorubicin), Lam 67R (Laminin 67R), NF-κB (nuclear factor kappa B), Bcl-xL (anti-apoptotic protein), Glutathione (GSH), Bax, Bcl-2 (antiapoptotic protein). nm (nanometer), μM (Micrometer), g (grams), ml (milliliter), μg (microgram).

NPs initially enter the body through different routes that can be inhalation, oral, ophthalmic, cutaneous, intratumoral (IT) or parenteral (intravenous) administration (Coutiño et al., 2017)(Coutiño et al., 2017). These routes of administration allow passage to the blood circulation, agglomerating in macrophages and subsequently dispersing to different organs (depending on the route of administration). However, cancer can be found far from the application site, so NPs have to travel in the circulatory system (Nichols & Bae, 2012). Therefore, once these NPs enter the body particularly in the bloodstream to go to their target sites, they can interact with some organs such as the lung, intestinal tract, liver, kidneys, spleen, among others (Ajdary et al., 2018; Hoet et al., 2004). In addition, NPs can be absorbed faster by the cells of the mononuclear phagocytic system (MPS) present in the kidney, stomach, heart, spleen, and liver considering sizes of 150-300 nm, and in the bone marrow with sizes of 30-150 nm. In this sense, the NPs-Inorganic before reaching their target sites could bio-distribute followed by crossing the epithelial barriers and traveling through the vascular bed and even escape from the circulation through fenestrations of the endothelial barrier (Coutiño et al., 2017; Gaumet et al., 2008; Llabot et al., 2008).

It should be noted that NPs can be initially degrade in the gastrointestinal tract by modulating pH, during peristalsis causing alterations in the physicochemical properties of NPs (Coradeghini et al., 2013; Hoet et al., 2004). In this process, due to the nanometric sizes of NPs they can be metabolized by the body penetrate the capillaries efficiently, and be absorbed by the reticuloendothelial system (RES) that will allow the elimination of foreign agents present in the circulation in prolonged times, which could derive in the accumulation of NPs in the liver or spleen that can later be eliminate from the body through urine, feces without causing toxic or harmful effects on the organs, as shown by some in vivo studies carried out in mice (Davatgaran-Taghipour et al., 2017; De Jong et al., 2013; Shukla et al., 2012; Yuan et al., 2018). Hence, NPs sizes $\leq$ 100 nm are absorbed into the bloodstream without contact with internal tissue surfaces (Devi & Sathishkumar, 2017) so they could be efficiently eliminated from the body, however, nanoparticles with sizes 20-100 nm are estimated to cause some liver damage (De Jong et al., 2013). But in an in vivo study carried out in 62 healthy volunteer patients to whom commercial solutions of AgNPs with sizes of 20 – 40 nm in diameter were administered orally, they did not present changes or damage in the metabolism and morphology of organs these effects were demonstrated by hematology and urine studies (Munger et al., 2014). For all the above it is important to consider that if there is no strategy for the administration of NPs, they could be distributed to various organs and cause toxicity to other cells (Alsalhi et al., 2016; Dehghanizade et al., 2018; Iqbal et al., 2020; Menon et al., 2012; Selim et al., 2020; Seo et al., 2014; Yallappa et al., 2016; Zhao, Chengbo, et al., 2018). In this sense, it is necessary to continue carrying out toxicological studies and research in vivo models concerning the toxic effects that NPs can cause in the human body to better understand how these nanomaterials act and the effects that could cause in the long term (Davatgaran-Taghipour et al., 2017).

Conclusion

Currently, there is a need to find an efficient cure for cancer, a situation that has led researchers to search for new treatments without destructive effects, improving the quality of life of the patient. Nanotechnologies focused on drug delivery in a targeted way through the of use nanoparticles with unique properties these nanostructures can improve their properties by functionalizing their surface with the doping of polyphenols that improve their efficiency and potentiate its effects. These combined materials are ideal for the treatment of cancer since they are biocompatible and have a better interaction with the biological systems of the human body reaching cellular levels, causing cell death measured by pro-apoptotic proteins and caspases. Therefore, there is a long way to go for the medical application of these polyphenol nanoformulations. However, this is one of the most suitable alternatives for treating cancer without causing side effects. It is expected that the future perspectives change the approach of doctors and scientists for polyphenol nanoformulations can have biomedical applications in the treatment of cancer and that they are a solution as a differential treatment in other multifactorial diseases. It is also expected that polyphenol nanoformulations can be accessible to patients and will have a positive socio-economic impact as by obtaining them through green synthesis, costs associated with the process are eliminated, and polyphenol-doped NPs can be successfully produced without contamination risks.

Acknowledgments

Acknowledgements BCCL thanks the National Council of Science and Technology (CONACYT) for the financial support during her Ph. D. studies. Financial support was received from Secretary of Agriculture and Rural Development (SAGARPA)-CONACyT through the project: "Obtenci´on, purificaci´on y escalado de compuestos de extractos bioactivos con valor industrial, obtenidos usando tecnolog´ıa avanzadas de extracci´on y a partir de cultivos, subproductos y recursos naturales poco valorados" SAGARPA-CONACyT 2015-4-266936. Thanks also NOVA book for allowing us to publish this information.

Conflict of Interest

The authors declare no conflict of interest.

References

Abbas, M., Saeed, F., Anjum, F. M., Afzaal, M., Tufail, T., Bashir, M. S., Ishtiaq, A., Hussain, S., & Suleria, H. A. R. (2017). Natural polyphenols: An overview. *International Journal of Food Properties*, *20*(8), 1689–1699. https://doi.org/10.1080/10942912.2016.1220393.

Abdullaeva, Z. (2017). Synthesis of Nanoparticles and Nanomaterials. In *Synthesis of Nanoparticles and Nanomaterials*. https://doi.org/10.1007/978-3-319-54075-7.

Ajdary, M., Moosavi, M. A., Rahmati, M., Falahati, M., Mahboubi, M., Mandegary, A., Jangjoo, S., Mohammadinejad, R., & Varma, R. S. (2018). Health concerns of various nanoparticles: A review of their in vitro and in vivo toxicity. *Nanomaterials*, *8*(9), 1–28. https://doi.org/10.3390/nano8090634.

Albulet, D., Florea, D. A., Boarca, B., Ditu, L. M., Chifiriuc, M. C., Grumezescu, A. M., & Andronescu, E. (2017). Nanotechnology for personalized medicine: Cancer research, diagnosis, and therapy. In *Nanostructures for Cancer Therapy*. Elsevier Inc. https://doi.org/10.1016/B978-0-323-46144-3.00001-5.

Ali, O. M., Bekhit, A. A., Khattab, S. N., Helmy, M. W., Abdel-Ghany, Y. S., Teleb, M., & Elzoghby, A. O. (2020). Synthesis of lactoferrin mesoporous silica nanoparticles for pemetrexed/ellagic acid synergistic breast cancer therapy. *Colloids and Surfaces B: Biointerfaces*, *188* (August 2019), 110824. https://doi.org/10.1016/j.colsurfb.2020.110824.

Alsalhi, M. S., Devanesan, S., Alfuraydi, A. A., Vishnubalaji, R., Munusamy, M. A., Murugan, K., Nicoletti, M., & Benelli, G. (2016). Green synthesis of silver nanoparticles using Pimpinella anisum seeds: Antimicrobial activity and cytotoxicity on human neonatal skin stromal cells and colon cancer cells. *International Journal of Nanomedicine*, *11*, 4439–4449. https://doi.org/10.2147/IJN.S113193.

Alshatwi, A., A., Athinarayanan, J., & Subbarayan, P. V. (2015a). Green synthesis of bimetallic Au@Pt nanostructures and their application for proliferation inhibition and apoptosis induction in human cervical cancer cell. *Journal of Materials Science: Materials in Medicine*, *26*(3), 2–9. https://doi.org/10.1007/s10856-015-5468-5.

Alshatwi, A., A., Athinarayanan, J., & Subbarayan, V. P. (2015b). Green synthesis of platinum nanoparticles that induce cell death and G2/M-phase cell cycle arrest in human cervical cancer cells. *Journal of Materials Science: Materials in Medicine*, *26*(1), 2–9. https://doi.org/10.1007/s10856-014-5330-1.

Badea, N., Craciun, M. M., Dragomir, A. S., Balas, M., Dinischiotu, A., Nistor, C., Gavan, C., & Ionita, D. (2020). Systems based on carbon nanotubes with potential in cancer therapy. *Materials Chemistry and Physics*, *241*(October 2019), 122435. https://doi.org/10.1016/j.matchemphys.2019.122435.

Bethu, M. S., Netala, V. R., Domdi, L., Tartte, V., & Janapala, V. R. (2018). Potential anticancer activity of biogenic silver nanoparticles using leaf extract of Rhynchosia suaveolens: an insight into the mechanism. *Artificial Cells, Nanomedicine and Biotechnology*, *46*(sup1), 104–114. https://doi.org/10.1080/21691401.2017.1414824.

Bodelón, G., Costas, C., Pérez-Juste, J., Pastoriza-Santos, I., & Liz-Marzán, L. M. (2017). Gold nanoparticles for regulation of cell function and behavior. *Nano Today*, *13*, 40–60. https://doi.org/10.1016/j.nantod.2016.12.014.

Bray, F., Ferlay, J., Soerjomataram, I., Siegel, R. L., Torre, L. A., & Jemal, A. (2018). Global cancer statistics 2018: GLOBOCAN estimates of incidence and mortality worldwide for 36 cancers in 185 countries. *CA: A Cancer Journal for Clinicians*, *68*(6), 394–424. https://doi.org/10.3322/caac.21492.

Buttacavoli, M., Albanese, N. N., Di Cara, G., Alduina, R., Faleri, C., Gallo, M., Pizzolanti, G., Gallo, G., Feo, S., Baldi, F., & Cancemi, P. (2018). Anticancer activity of biogenerated silver nanoparticles: An integrated proteomic investigation. *Oncotarget*, *9*(11), 9685–9705. https://doi.org/10.18632/oncotarget.23859.

Castro-Acosta, M. L. (2019). Polifenoles: compuestos bioactivos con efectos benéficos en la prevención de diabetes tipo 2. *REDCieN - Ciencia y Nutrición*, *1*(3), 13–18.

Choudhari, A. S., Mandave, P. C., Deshpande, M., Ranjekar, P., & Prakash, O. (2020). Phytochemicals in cancer treatment: From preclinical studies to clinical practice. *Frontiers in Pharmacology*, *10*(January), 1–17. https://doi.org/10.3389/fphar.2019.01614.

Chugh, H., Sood, D., Chandra, I., Tomar, V., Dhawan, G., & Chandra, R. (2018). Role of gold and silver nanoparticles in cancer nano-medicine. *Artificial Cells, Nanomedicine and Biotechnology*, *46*(sup1), 1210–1220. https://doi.org/10.1080/21691401.2018.1449118.

Coradeghini, R., Gioria, S., García, C. P., Nativo, P., Franchini, F., Gilliland, D., Ponti, J., & Rossi, F. (2013). Size-dependent toxicity and cell interaction mechanisms of gold nanoparticles on mouse fibroblasts. *Toxicology Letters*, *217*(3), 205–216. https://doi.org/10.1016/j.toxlet.2012.11.022.

Coutiño, E. M. del R., Ávila, L. L., & Arroyo, H. O. (2017). Las nanopartículas de plata: mecanismos de entrada, toxicidad y estrés oxidativo. *Revista de Educación Bioquímica*, *36*(2), 39–54. [Silver nanoparticles: mechanisms of entry, toxicity and oxidative stress. *Journal of Biochemical Education*]

Daneshmehr, S. (2015). Carbon Nanotubes for Delivery of Quercetin as Anticancer Drug: Theoretical Study. *Procedia Materials Science*, *11*(2014), 131–136. https://doi.org/10.1016/j.mspro.2015.11.128.

Davatgaran-Taghipour, Y., Masoomzadeh, S., Farzaei, M. H., Bahramsoltani, R., Karimi-Soureh, Z., Rahimi, R., & Abdollahi, M. (2017). Polyphenol nanoformulations for cancer therapy: Experimental evidence and clinical perspective. *International Journal of Nanomedicine*, *12*, 2689–2702. https://doi.org/10.2147/IJN.S131973.

De Jong, W. H., Van Der Ven, L. T. M., Sleijffers, A., Park, M. V. D. Z., Jansen, E. H. J. M., Van Loveren, H., & Vandebriel, R. J. (2013). Systemic and immunotoxicity of silver nanoparticles in an intravenous 28 days repeated dose toxicity study in rats. *Biomaterials*, *34*(33), 8333–8343. https://doi.org/10.1016/j.biomaterials.2013.06.048.

De Matteis, V. (2017). Exposure to inorganic nanoparticles: Routes of entry, immune response, biodistribution and in vitro/In vivo toxicity evaluation. *Toxics*, *5*(4). https://doi.org/10.3390/toxics5040029.

Dehghanizade, S., Arasteh, J., & Mirzaie, A. (2018). Green synthesis of silver nanoparticles using Anthemis atropatana extract: characterization and in vitro biological activities. *Artificial Cells, Nanomedicine and Biotechnology*, *46*(1), 160–168. https://doi.org/10.1080/21691401.2017.1304402.

Devanesan, S., AlSalhi, M. S., Balaji, R. V., Ranjitsingh, A. J. A., Ahamed, A., Alfuraydi, A. A., AlQahtani, F. Y., Aleanizy, F. S., & Othman, A. H. (2018). Antimicrobial and Cytotoxicity Effects of Synthesized Silver Nanoparticles from Punica granatum Peel Extract. *Nanoscale Research Letters*, *13*(1), 1–10. https://doi.org/10.1186/s11671-018-2731-y.

Devi, G. K., & Sathishkumar, K. (2017). Synthesis of gold and silver nanoparticles using Mukia maderaspatna plant extract and its anticancer activity. *IET Nanobiotechnology*, *11*(2), 143–151. https://doi.org/10.1049/iet-nbt.2015.0054.

Fariq, A., Khan, T., & Yasmin, A. (2017). Microbial synthesis of nanoparticles and their potential applications in biomedicine. *Journal of Applied Biomedicine*, *15*(4), 241–248. https://doi.org/10.1016/j.jab.2017.03.004.

Filip, G. A., Moldovan, B., Baldea, I., Olteanu, D., Suharoschi, R., Decea, N., Cismaru, C. M., Gal, E., Cenariu, M., Clichici, S., & David, L. (2019). UV-light mediated green synthesis of silver and gold nanoparticles using Cornelian cherry fruit extract and their comparative effects in experimental inflammation. *Journal of Photochemistry and Photobiology B: Biology*, *191*(December 2018), 26–37. https://doi.org/10.1016/j.jphotobiol. 2018.12.006.

Friehs, E., AlSalka, Y., Jonczyk, R., Lavrentieva, A., Jochums, A., Walter, J. G., Stahl, F., Scheper, T., & Bahnemann, D. (2016). Toxicity, phototoxicity and biocidal activity of nanoparticles employed in photocatalysis. *Journal of Photochemistry and Photobiology C: Photochemistry Reviews*, *29*, 1–28. https://doi.org/10.1016/j.jphotochemrev.2016.09.001.

Gaumet, M., Vargas, A., Gurny, R., & Delie, F. (2008). Nanoparticles for drug delivery: The need for precision in reporting particle size parameters. *European Journal of Pharmaceutics and Biopharmaceutics*, *69*(1), 1–9. https://doi.org/10.1016/j.ejpb.2007.08.001.

Gómez-Garzón, M. (2018). Nanomateriales, Nanopartículas y Síntesis verde. *Revista Repertorio de Medicina y Cirugía*, *27*(2), 75–80. https://doi.org/10. 31260/repertmedcir.v27.n2.2018.191. [Nanomaterials, Nanoparticles and Green Synthesis. *Magazine Repertory of Medicine and Surgery*]

Govender, R., Phulukdaree, A., Gengan, R. M., Anand, K., & Chuturgoon, A. A. (2013). Silver nanoparticles of Albizia adianthifolia: The induction of apoptosis in human lung carcinoma cell line. *Journal of Nanobiotechnology*, *11*(1), 1–9. https://doi.org/10.1186/1477-3155-11-5.

Guía-García, J. L., Charles-Rodríguez, A. V., López-Romero, J. C., Torres-Moreno, H., Genisheva, Z., Robledo-Olivo, A., Reyes-Valdés, H. M., Ramírez-Godina, F., García-Osuna, H. T., & Flores-López, M. L. (2021). Phenolic Composition and Biological Properties of Rhus microphylla and Myrtillocactus geometrizans Fruit Extracts Jorge. *Plants*, *10*, 1–15.

Gurunathan, S., Han, J. W., Park, J. H., Kim, E., Choi, Y. J., Kwon, D. N., & Kim, J. H. (2015). Reduced graphene oxide-silver nanoparticle nanocomposite: A potential anticancer nanotherapy. *International Journal of Nanomedicine*, *10*, 6257–6276. https://doi.org/10.2147/IJN.S92449.

He, Y., Li, X., Zheng, Y., Wang, Z., Ma, Z., Yang, Q., Yao, B., Zhao, Y., & Zhang, H. (2018). A green approach for synthesizing silver nanoparticles, and their antibacterial and cytotoxic activities. *New Journal of Chemistry*, *42*(4), 2882–2888. https://doi.org/10.1039/c7nj04224h.

Hoet, P. H. M., Brüske-Hohlfeld, I., & Salata, O. V. (2004). Nanoparticles - Known and unknown health risks. *Journal of Nanobiotechnology*, *2*, 1–15. https://doi.org/10.1186/1477-3155-2-12.

Hsieh, D. S., Lu, H. C., Chen, C. C., Wu, C. J., & Yeh, M. K. (2012). The preparation and characterization of gold-conjugated polyphenol nanoparticles as a novel delivery system. *International Journal of Nanomedicine*, *7*, 1623–1633. https://doi.org/10.2147/IJN.S30060.

Iqbal, J., Abbasi, B. A., Ahmad, R., Mahmood, T., Ali, B., Khalil, A. T., Kanwal, S., Shah, S. A., Alam, M. M., Badshah, H., & Munir, A. (2018). Nanomedicines for developing cancer nanotherapeutics: from benchtop to bedside and beyond. *Applied Microbiology and Biotechnology*, *102*(22), 9449–9470. https://doi.org/10.1007/s00253-018-9352-3.

Iqbal, J., Abbasi, B. A., Munir, A., Uddin, S., Kanwal, S., & Mahmood, T. (2020). Facile green synthesis approach for the production of chromium oxide nanoparticles and their different in vitro biological activities. *Microscopy Research and Technique*, *83*(6), 706–719. https://doi.org/10.1002/jemt.23460.

Kabera, J. N., Semana, E., Mussa, A. R., & He, X. (2014). Plant Secondary Metabolites: Biosynthesis, Classification, Function and Pharmacological Properties. *Journal of Pharmacy and Pharmacology*, *2*(January), 377–392.

Khan, I., Saeed, K., & Khan, I. (2019). Nanoparticles : Properties, applications and toxicities. *Arabian Journal of Chemistry*, *12*(7), 908–931. https://doi.org/10.1016/j.arabjc.2017.05.011.

Lee, K. X., Shameli, K., Mohamad, S. E., Yew, Y. P., Isa, E. D. M., Yap, H. Y., Lim, W. L., & Teow, S. Y. (2019). Bio-mediated synthesis and characterisation of silver nanocarrier, and its potent anticancer action. *Nanomaterials*, *9*(10), 1–19. https://doi.org/10.3390/nano9101423.

Liang, J., Li, F., Fang, Y., Yang, W., An, X., Zhao, L., Xin, Z., Cao, L., & Hu, Q. (2011). Synthesis, characterization and cytotoxicity studies of chitosan-coated tea polyphenols nanoparticles. *Colloids and Surfaces B: Biointerfaces*, *82*(2), 297–301. https://doi.org/10.1016/j.colsurfb.2010.08.045.

Llabot, J. M., Palma, S. D., & Allemandi, D. A. (2008). *Nanopartículas poliméricas sólidas*. 40–47. [*Solid polymer nanoparticles*]

López-Gomez., A. (2017). Repertorio de Medicina y Cirugía Nanomedicine and its impact on medical practice. *Nanomedicina y Su Impacto En La Práctica Médica*, *6*(3), 129–130. [Repertory of Medicine and Surgery Nanomedicine and its impact on medical practice. *Nanomedicine and Its Impact on Medical Practice*]

Martínez, E. M. M., Sandate-Flores, L., Rodríguez-Rodríguez, J., Rostro-Alanis, M., Parra-Arroyo, L., Antunes-Ricardo, M., Serna-Saldívar, S. O., Iqbal, H. M. N., & Parra-Saldívar, R. (2021). Underutilized mexican plants: Screening of antioxidant and antiproliferative properties of mexican cactus fruit juices. *Plants*, *10*(2), 1–13. https://doi.org/10.3390/plants10020368.

Menon, D., Basanth, A., Retnakumari, A., Manzoor, K., & Nair, S. V. (2012). Green synthesis of biocompatible gold nanocrystals with tunable surface plasmon resonance using garlic phytochemicals. *Journal of Biomedical Nanotechnology*, *8*(6), 901–911. https://doi.org/10.1166/jbn.2012.1455.

Mercado-Mercado, G., Carrillo, L. de la R., Wall-Medrano, A., Díaz, J. A. L., & Álvarez-Parrilla, E. (2013). Compuestos polifenólicos y capacidad antioxidante de especias típicas consumidas en México. *Nutricion Hospitalaria*, *28*(1), 36–46. https://doi.org/10.3305/nh.2013.28.1.6298. [Polyphenolic compounds and antioxidant capacity of typical spices consumed in Mexico. *Hospital Nutrition*]

Mirza, A. Z., & Siddiqui, F. A. (2014). Nanomedicine and drug delivery: a mini review. *International Nano Letters*, *4*(1). https://doi.org/10.1007/s40089-014-0094-7.

Montané, X., Bajek, A., Roszkowski, K., Montornés, J. M., Giamberini, M., Roszkowski, S., Kowalczyk, O., Garcia-Valls, R., & Tylkowski, B. (2020). Encapsulation for cancer therapy. *Molecules*, *25*(7), 1–25. https://doi.org/10.3390/molecules25071605.

Mukherjee, Sudeshna, Ghosh, S., Das, D. K., Chakraborty, P., Choudhury, S., Gupta, P., Adhikary, A., Dey, S., & Chattopadhyay, S. (2015). Gold-conjugated green tea nanoparticles for enhanced anti-tumor activities and hepatoprotection - synthesis, characterization and in vitro evaluation. *Journal of Nutritional Biochemistry*, *26*(11), 1283–1297. https://doi.org/10.1016/j.jnutbio.2015.06.003.

Munger, M. A., Radwanski, P., Hadlock, G. C., Stoddard, G., Shaaban, A., Falconer, J., Grainger, D. W., & Deering-Rice, C. E. (2014). In vivo human time-exposure study of orally dosed commercial silver nanoparticles. *Nanomedicine: Nanotechnology, Biology, and Medicine*, *10*(1), 1–9. https://doi.org/10.1016/j.nano.2013.06.010.

Nichols, J. W., & Bae, Y. H. (2012). Odyssey of a cancer nanoparticle: From injection site to site of action. *Nano Today*, *7*(6), 606–618. https://doi.org/10.1016/j.nantod.2012.10.010.

Noah, N. (2019). Green synthesis: Characterization and application of silver and gold nanoparticles. In *Green Synthesis, Characterization and Applications of*

Nanoparticles. Elsevier Inc. https://doi.org/10.1016/b978-0-08-102579-6.00006-x.

Oh, K. H., Soshnikova, V., Markus, J., Kim, Y. J., Lee, S. C., Singh, P., Castro-Aceituno, V., Ahn, S., Kim, D. H., Shim, Y. J., Kim, Y. J., & Yang, D. C. (2018). Biosynthesized gold and silver nanoparticles by aqueous fruit extract of Chaenomeles sinensis and screening of their biomedical activities. *Artificial Cells, Nanomedicine and Biotechnology*, *46*(3), 599–606. https://doi.org/10.1080/21691401.2017.1332636.

Ovais, M., Khalil, A. T., Raza, A., Khan, M. A., Ahmad, I., Islam, N. U., Saravanan, M., Ubaid, M. F., Ali, M., & Shinwari, Z. K. (2016). Green synthesis of silver nanoparticles via plant extracts: Beginning a new era in cancer theranostics. *Nanomedicine*, *12*(23), 3157–3177. https://doi.org/10.2217/nnm-2016-0279.

Ozturk, B., Parkinson, C., & Gonzalez-Miquel, M. (2018). Extraction of polyphenolic antioxidants from orange peel waste using deep eutectic solvents. *Separation and Purification Technology*, *206*(March), 1–13. https://doi.org/10.1016/j.seppur.2018.05.052.

Papuc, C., Goran, G. V., Predescu, C. N., Nicorescu, V., & Stefan, G. (2017). Plant Polyphenols as Antioxidant and Antibacterial Agents for Shelf-Life Extension of Meat and Meat Products: Classification, Structures, Sources, and Action Mechanisms. *Comprehensive Reviews in Food Science and Food Safety*, *16*(6), 1243–1268. https://doi.org/10.1111/1541-4337.12298.

Peñarrieta, J. M., Tejeda, L., Mollinedo, P., Vila, J. L. ., & Bravo, J. A. (2014). Phenolic compounds in food. *Bolivian Journal of Chemestry*, *31*(2), 68–81.

Qian, L., Su, W., Wang, Y., Dang, M., Zhang, W., & Wang, C. (2019). Synthesis and characterization of gold nanoparticles from aqueous leaf extract of Alternanthera sessilis and its anticancer activity on cervical cancer cells (HeLa). *Artificial Cells, Nanomedicine and Biotechnology*, *47*(1), 1173–1180. https://doi.org/10.1080/21691401.2018.1549064.

Raghunandan, D., Ravishankar, B., Sharanbasava, G., Mahesh, D. B., Harsoor, V., Yalagatti, M. S., Bhagawanraju, M., & Venkataraman, A. (2011). Anti-cancer studies of noble metal nanoparticles synthesized using different plant extracts. *Cancer Nanotechnology*, *2*(1–6), 57–65. https://doi.org/10.1007/s12645-011-0014-8.

Rasheed, T., Bilal, M., Iqbal, H. M. N., & Li, C. (2017). Green biosynthesis of silver nanoparticles using leaves extract of Artemisia vulgaris and their potential biomedical applications. *Colloids and Surfaces B: Biointerfaces*, *158*, 408–415. https://doi.org/10.1016/j.colsurfb.2017.07.020.

Raymundo-Pereira, P. A., Silva, T. A., Zapp, E., Brondani, D., Bergamini, M. F., Marcolino, L. H., Banks, C. E., Oliveira, O. N., Janegitz, B. C., & Fatibello-filho, O. (2020). Polyphenol oxidase-based electrochemical biosensors : A review. *Analytica Chimica Acta*, *xxxx*. https://doi.org/10.1016/j.aca.2020.07.055.

Reyes-Farias, M., & Carrasco-Pozo, C. (2019). The anti-cancer effect of quercetin: Molecular implications in cancer metabolism. *International Journal of Molecular Sciences*, *20*(13), 1–19. https://doi.org/10.3390/ijms20123177.

Riahi-Chebbi, I., Souid, S., Othman, H., Haoues, M., Karoui, H., Morel, A., Srairi-Abid, N., Essafi, M., & Essafi-Benkhadir, K. (2019). The Phenolic compound Kaempferol overcomes 5-fluorouracil resistance in human resistant LS174 colon cancer cells. *Scientific Reports*, *9*(1), 1–20. https://doi.org/10.1038/s41598-018-36808-z.

Rojas, G. N. (2012). *Evaluacion de fenolicos totales y capacidad antioxidante en la pulpa concentrada de zarzamora (Rubus sp) en dos estadios de madurez*. Universidad Nacional del Centro del Peru. [*Evaluation of total phenolics and antioxidant capacity in the concentrated pulp of blackberry (Rubus sp) in two stages of maturity*.]

Sánchez, C. N. (2013). Conociendo y comprendiendo la célula cancerosa: Fisiopatología del cáncer. *Revista Médica Clínica Las Condes*, *24*(4), 553–562. https://doi.org/10.1016/s0716-8640(13)70659-x. [Knowing and understanding the cancer cell: Pathophysiology of cancer. *Las Condes Clinical Medical Journal*]

Sanna, V., Pala, N., DessÌ, G., Manconi, P., Mariani, A., Dedola, S., Rassu, M., Crosio, C., Iaccarino, C., & Sechi, M. (2014). Single-step green synthesis and characterization of gold-conjugated polyphenol nanoparticles with antioxidant and biological activities. *International Journal of Nanomedicine*, *9*(1), 4935–4951. https://doi.org/10.2147/IJN.S70648.

Selim, Y. A., Azb, M. A., Ragab, I., & H. M. Abd El-Azim, M. (2020). Green Synthesis of Zinc Oxide Nanoparticles Using Aqueous Extract of Deverra tortuosa and their Cytotoxic Activities. *Scientific Reports*, *10*(1), 1–9. https://doi.org/10.1038/s41598-020-60541-1.

Seo, Y., Hwang, J., Kim, J., Jeong, Y., Hwang, M. P., & Choi, J. (2014). Antibacterial activity and cytotoxicity of multi-walled carbon nanotubes decorated with silver nanoparticles. *International Journal of Nanomedicine*, *9*(1), 4621–4629. https://doi.org/10.2147/IJN.S69561.

Shukla, R., Chanda, N., Zambre, A., Upendran, A., Katti, K., Kulkarni, R. R., Nune, S. K., Casteel, S. W., Smith, C. J., Vimal, J., Boote, E., Robertson, J. D., Kan, P., Engelbrecht, H., Watkinson, L. D., Carmack, T. L., Lever, J. R., Cutler, C. S., Caldwell, C., … Katti, K. V. (2012). Laminin receptor specific therapeutic gold nanoparticles (198AuNP-EGCg) show efficacy in treating prostate cancer. *Proceedings of the National Academy of Sciences of the*

United States of America, *109*(31), 12426–12431. https://doi.org/10.1073/pnas.1121174109.

Siddiqui, M. A., Wahab, R., Ahmad, J., Farshori, N. N., Musarrat, J., & Al-Khedhairy, A. A. (2017). Evaluation of cytotoxic responses of raw and functionalized multi-walled carbon nanotubes in human breast cancer (MCF-7) cells. *Vacuum*, *146*, 578–585. https://doi.org/10.1016/j.vacuum.2017.05.022.

Singh, H., Du, J., & Yi, T. H. (2017). Green and rapid synthesis of silver nanoparticles using Borago officinalis leaf extract: anticancer and antibacterial activities. *Artificial Cells, Nanomedicine and Biotechnology*, *45*(7), 1310–1316. https://doi.org/10.1080/21691401.2016.1228663.

Son, K. H., Hong, J. H., & Lee, J. W. (2016). Carbon nanotubes as cancer therapeutic carriers and mediators. *International Journal of Nanomedicine*, *11*, 5163–5185. https://doi.org/10.2147/IJN.S112660.

Tang, Si-min., Deng, X., Zhou, J., Li, Q., Ge, X., & Miao, L. (2020). Biomedicine & Pharmacotherapy Pharmacological basis and new insights of quercetin action in respect to its anti-cancer e ff ects. *Biomedicine & Pharmacotherapy*, *121*(July 2019), 109604. https://doi.org/10.1016/j.biopha.2019.109604.

Tang, W., Li, W., Yang, Y., Lin, X., Wang, L., Li, C., & Yang, R. (2021). Phenolic compounds profile and antioxidant capacity of pitahaya fruit peel from two red-skinned species (*Hylocereus polyrhizus* and *Hylocereus undatus*). *Foods*, *10*(6). https://doi.org/10.3390/foods10061183.

Thipe, V. C., Amiri, K. P., Bloebaum, P., Karikachery, A. R., Khoobchandani, M., Katti, K. K., Jurisson, S. S., & Katti, K. V. (2019). Development of resveratrol-conjugated gold nanoparticles: Interrelationship of increased resveratrol corona on anti-tumor efficacy against breast, pancreatic and prostate cancers. *International Journal of Nanomedicine*, *14*, 4413–4428. https://doi.org/10.2147/IJN.S204443.

Trung, L. Q., & An, D. T. T. (2018). Is Resveratrol a cancer immunomodulatory molecule? *Frontiers in Pharmacology*, *9*(NOV), 2010–2013. https://doi.org/10.3389/fphar.2018.01255.

Vervandier-Fasseur, D., & Latruffe, N. (2019). The potential use of resveratrol for cancer prevention. *Molecules*, *24*(24). https://doi.org/10.3390/molecules24244506.

Wang, Dandan., Markus, J., Wang, C., Kim, Y. J., Mathiyalagan, R., Aceituno, V. C., Ahn, S., & Yang, D. C. (2017). Green synthesis of gold and silver nanoparticles using aqueous extract of *Cibotium barometz* root. *Artificial Cells, Nanomedicine and Biotechnology*, *45*(8), 1548–1555. https://doi.org/10.1080/21691401.2016.1260580.

Wang, T. yang, Li, Q., & Bi, K. shun. (2018). Bioactive flavonoids in medicinal plants: Structure, activity and biological fate. *Asian Journal of*

Pharmaceutical Sciences, *13*(1), 12–23. https://doi.org/10.1016/j.ajps.2017.08.004.

Wang, Xueni., Yang, Y., An, Y., & Fang, G. (2019). Biomedicine & Pharmacotherapy The mechanism of anticancer action and potential clinical use of kaempferol in the treatment of breast cancer. *Biomedicine & Pharmacotherapy*, *117*(May), 109086. https://doi.org/10.1016/j.biopha.2019.109086.

Wang, Zhi., Que, B., Gan, J., Guo, H., Chen, Q., Zheng, L., Marraiki, N., Elgorban, A. M., & Zhang, Y. (2020). Zinc oxide nanoparticles synthesized from Fraxinus rhynchophylla extract by green route method attenuates the chemical and heat induced neurogenic and inflammatory pain models in mice. *Journal of Photochemistry and Photobiology B: Biology*, *202*(October 2019), 111668. https://doi.org/10.1016/j.jphotobiol.2019.111668.

Wei, L., Lu, J., Xu, H., Patel, A., Chen, Z. S., & Chen, G. (2015). Silver nanoparticles: Synthesis, properties, and therapeutic applications. *Drug Discovery Today*, *20*(5), 595–601. https://doi.org/10.1016/j.drudis.2014.11.014.

Williamson, G. (2017). The role of polyphenols in modern nutrition. *Nutrition Bulletin*, *42*(3), 226–235. https://doi.org/10.1111/nbu.12278.

Yallappa, S., Manjanna, J., Dhananjaya, B. L., Vishwanatha, U., Ravishankar, B., Gururaj, H., Niranjana, P., & Hungund, B. S. (2016). Phytochemically Functionalized Cu and Ag Nanoparticles Embedded in MWCNTs for Enhanced Antimicrobial and Anticancer Properties. *Nano-Micro Letters*, *8*(2), 120–130. https://doi.org/10.1007/s40820-015-0066-0.

Yi, X., Zuo, J., Tan, C., Xian, S., Luo, C., Chen, S., Yu, L., & Luo, Y. (2016). *kaempferol, a flavonoid compound from gynura medica induced apoptosis and growth inhibition in mcf-7 breast cancer cell. 13*, 210–215. https://doi.org/10.21010/ajtcam.v13i4.27.

Yoosefian, M., Sabaei, S., & Etminan, N. (2019). Encapsulation efficiency of single-walled carbon nanotube for Ifosfamide anti-cancer drug. *Computers in Biology and Medicine*, *114*(May), 103433. https://doi.org/10.1016/j.compbiomed.2019.103433.

Yuan, Xiaoming, He, Y., Zhou, G., Li, X., Feng, A., & Zheng, W. (2018). Target challenging-cancer drug delivery to gastric cancer tissues with a fucose graft epigallocatechin-3-gallate-gold particles nanocomposite approach. *Journal of Photochemistry and Photobiology B: Biology*, *183*(April), 147–153. https://doi.org/10.1016/j.jphotobiol.2018.04.026.

Yuan, Y. G., Peng, Q. L., & Ggurunathan, S. (2017). Silver nanoparticles enhance the apoptotic potential of gemcitabine in human ovarian cancer cells: Combination therapy for effective cancer treatment. *International Journal of Nanomedicine*, *12*, 6487–6502. https://doi.org/10.2147/IJN.S135482.

Zhang, D., Zhang, J., Zeng, J., Li, Z., Zuo, H., Huang, C., & Zhao, X. (2019). Nano-gold loaded with resveratrol enhance the anti-hepatoma effect of resveratrol in vitro and in vivo. *Journal of Biomedical Nanotechnology*, *15*(2), 288–300. https://doi.org/10.1166/jbn.2019.2682.

Zhao, Chengbo, Zhang, X., & Zheng, Y. (2018). Biosynthesis of polyphenols functionalized ZnO nanoparticles: Characterization and their effect on human pancreatic cancer cell line. *Journal of Photochemistry and Photobiology B: Biology*, *183*(17), 142–146. https://doi.org/10.1016/j.jphotobiol.2018.04.031.

Zhao, Xia-Mei, Cai, Z. C., Zhu, B. J., & Zhang, Z. Q. (2018). The Apoptosis Effect on Liver Cancer Cells of Gold Nanoparticles Modified with Lithocholic Acid. *Nanoscale Research Letters*, *13*, 0–7. https://doi.org/10.1186/s11671-018-2653-8.

Zhu, B., Li, Y., Lin, Z., Zhao, M., Xu, T., Wang, C., & Deng, N. (2016). Silver Nanoparticles Induce HePG-2 Cells Apoptosis Through ROS-Mediated Signaling Pathways. *Nanoscale Research Letters*, *11*(1), 1–8. https://doi.org/10.1186/s11671-016-1419-4.

Chapter 3

Functional Activity of Purslane (*Portulaca oleracea* L.) and Its Use in the Food Industry

Carolina González-Rodríguez, Adriana Carolina Flores-Gallegos, Juan Alberto Ascacio-Valdés, Mónica Lizeth Chávez-González, Lissethe Palomo-Ligas and Raúl Rodríguez-Herrera*

Food Research Department, School of Chemistry,
Universidad Autónoma de Coahuila, Saltillo, Coahuila, México

Abstract

Functional foods contain nutritional and biologically active compounds which provide health benefits, this type of food can be obtained naturally from different plants. One of these plants is purslane (*Portulaca oleracea* L.) which is distributed worldwide, mainly in England, Saudi Arabia, India, China, and some African countries, it is even well distributed throughout Mexico. In the last country, it is identified as an ancestral and traditional food, although in recent years, it is less consumed, mainly with the development of highly processed foods and those of fast preparation, so the nutritional properties and health benefits that this plant can provide to the human body are little known. On the other hand, obesity and

* Corresponding Author's E-mail: raul.rodriguez@uadec.edu.mx.

In: Food Processing
Editors: Myriam Huijs
ISBN: 978-1-68507-570-5

cardiovascular diseases are some of the main health problems in different countries. Thus, it is essential to determine the impact of different foods on health, in order to develop an ingredient or functional food that helps to solve or prevent some of the health problems mentioned above. The objective of this document was the search and discuss recent literature on *Portulaca oleracea*. It was possible to identify the phytochemicals present in purslane, its health benefits, and its applications in the food industry.

Keywords: *Portulaca oleracea*, phytochemicals, foods, functional foods

Introduction

Purslane (*Portulaca oleracea*) is an annual herbaceous plant that belongs to the *Portulacaceae* family, which can reach 40 cm in height and has yellow flowers. This plant is distributed around the world due to its adaptation to different climates and adverse conditions, although in some places it is considered a weed (Uddin et al., 2012). There is little information about the origin of purslane, however, in traditional Greek medicine, it was used as a remedy for constipation and to treat inflammation in the urinary system. On the other hand, in India, different properties have been attributed to this plant to treat liver problems, while in the Iberian Peninsula it has traditionally been used in different ways, regularly as an infusion, to treat inflamed hemorrhoids, insect bites, abscesses, sores, diarrhea, intestinal bleeding, and fluid retention among others (Ethnobotany, 2021). In Mexico, the main use of purslane is as food either as a cooked or boiled vegetable in soups and salads, mainly in the central area of the country and in some ethnic areas in the Mexican Republic (National Seed Inspection and Certification Service., 2017).

Portulaca oleracea has been studied from a pharmacological point of view due to its antifungal, antioxidant, anti-inflammatory, and antimicrobial properties (Desta et al., 2020). The objective of this research was to collect recent information about *P. oleracea* phytochemicals, its nutritional and pharmacological activities, and purslane potential for the development of functional foods, in addition to socializing the scientific information about *P. oleracea*. This plant is low-valued in modern society, but for many centuries, humanity has used it, as a dietary ingredient or as a medical treatment, and this knowledge has passed from generation to generation.

Analytical Framework

Literature Review

Portulaca Genus (Portulacaceae)

This genus is characterized by having annual, biennial, or perennial plants, herbaceous, sometimes fleshy, which can be prostrate or semi-prostrate, with simple or branched stems from their base. They are made up of simple leaves which can be alternate or sub-posed, with a flattened or rounded blade. *Portulaca* plants have terminal inflorescences, with flowers, this genus comprises about 50 species distributed throughout the tropical and subtropical areas around the world, mainly in Africa and South America (De Lorenzo-Cáceres, 2021). Some of the *Portulaca* species are: *P. amilis, P. grandiflora, P. pilosa, P. quadrifida, P. califórnica, P. oleracea*, among others (Ocampo and Columbus, 2012).

Taxonomy

Portulaca oleraceae L, traditionally known in Mexico as purslane, belongs to the Portulacaceae family. This family has about 20-50 genera with approximately 400 species. The meaning of the name Portulaca refers to milk, which comes from the Latin name "lac" because this plant contains a milky juice (Iranshahy et al., 2017). It is an annual, herbaceous, succulent plant and has a sloping or erect growth habit. It has cylindrical stems of a reddish color, with radially extending branches, the roots can form out of their usual place at the stem nodes. Its leaves can be alternate and can even be opposite, egg-shaped or wedge-shaped to spatulate, approximately 0.5 to 3 cm long and 0.2 to 1.5 cm wide (Calderón de Rzedowski, 2001).

Distribution

The purslane *Portulaca oleraceae L* is present in India, Saudi Arabia, China, England, Africa, and Mexico. It has been reported in many other countries, and in some of them, it is considered a weed. Purslane cropping in Mexico began in the 1950s, mainly in the chinampera area of Mexico City, where currently, some types have been domesticated by the same producers (Martínez-López et al., 2020). In this country, it is possible to find purslane in wild areas and as a weed in Chiapas, Durango, Guanajuato, Guerrero, Hidalgo, Jalisco, State of Mexico, Michoacán, Morelos, Nuevo León, Oaxaca, Puebla, Querétaro, Quintana Roo, San Luis Potosí, Sinaloa, Tabasco, Tamaulipas, Tlaxcala, Veracruz, Yucatán, and Zacatecas States.

Cropping

Portulaca oleracea cropping can be performed in different ways. Purslane requires good soil and water conditions, and it is convenient to plant it in late spring or early summer, as it develops better on infertile and very well-watered land, since these conditions have a positive impact on the culinary characteristics of purslane, making it juicier (Berrones-Morales et al., 2007). Also, it is grown on organic matter-rich soils with a pH range from 5.5 to 5.8. This plant is very sensitive to high concentrations of salts and can be cultivated in areas with tropical, subtropical, temperate, and cold climates, the optimal temperature for this plant can range between 25.5°C and 26.5°C (SIAP, 2019). Other cropping ways of purslane are floating and hydroponic systems, which are promising for this plant because they merge all of the benefits that soilless cultivation offers higher plant density, and yield, better nutritional management, faster production, and cleaner cropping in comparison to traditional cropping (Fontana et al., 2006). An advantage of purslane is its adaptation to low-oxygen content and development of aerenchyma tissues, in environments with low-oxygen content, plant quality improved because nitrates content decreased, and the number of bioactive compounds increased in comparison to plants with higher oxygen content (Lara et al., 2011).

Applications

Since ancient times, this plant has been used as a medicinal and edible plant. Currently, it is still consumed as food in different countries, although in others, it is considered a weed and has become a low-valued or unimportant crop with a lack of care (Egea-Gilabert et al., 2014). Purslane has an acidic or slightly bitter taste and can be eaten as a salad, soups, and stews (Nemzer et al., 2020). Leaves and stems of this succulent plant can be consumed as vegetables in three variants: raw, cooked, or pickled in salads, although lately purslane has been considered a functional food due to its high nutritional value (Petropoulos et al., 2016). On the other hand, it has been found that purslane contains abundant compounds such as antioxidants, vitamins, omega-3 fatty acids, essential amino acids, and multiple minerals, mainly potassium (Gonnella et al., 2010). Mera-Ovando et al., (2010), mention that, in the medicinal area, purslane is recommended to treat digestive problems, intestinal infections, constipation, even parasites. On the other hand, it has been mentioned that purslane leaves and roots extracts have different applications in agriculture since they show a high potential to release compounds into the environment (especially phenolic acids, ferulic acid, chlorogenic acid, among others) through roots exudation, leaching, and decomposition of *P. oleracea* residues in the soil. In addition, some of its chemical components may have applications as growth regulators and/or natural herbicides (El-Shora and Abd El-Gawad, 2015).

Nutritional Composition

Purslane has a high content of minerals and carbohydrates, in addition to a considerable percentage of proteins (23.4-25.9%) and low caloric contribution (16 calories) (Table 1). On the other hand, it contains important calcium values, which is one of the main minerals in the human body, because it has a regulator effect on blood pressure, increases bone mineral density, prevents muscle contraction, and also intervenes during nerve signals (Pu et al., 2016). This plant contains high levels of potassium, this mineral is needed for its function in the nervous system, it is also a channel for nutrients to reach the cells and expel waste from them and forms the sodium-potassium pump that determines the electrical voltage of the factors that can cause potassium deficiency (Elcinto, 2000).

Table 1. Nutritional composition of purslane leaves (adapted from Román-Cortés et al., 2018)

Nutrient	**g/100 g Dry weight**
Moisture	1.3
Ash	17.9
Crude protein	25.9
Crude fiber	12.3
Carbohydrates	52.1
Lipids	4.1
Fatty acids	3.28
Minerals	mg/100 g Fresh weight
Nitrogen	1 651.3
Phosphor	329.5
Magnesium	1 121.5
Potassium	1 230.7
Calcium	512.7
Sodium	158.9
Iron	9.3
Manganese	3.8
Copper	1.2
Boron	5.3
Zinc	3.3

On the other hand, the phosphorus content is very important in purslane, since it is an essential nutrient for energy production, it acts as a signaling molecule and induces complex physiological responses, the role of this element is very important in the renal and cardiovascular systems, as well as in the brain (Bird and Eskin, 2021). Iron is found in high levels in *P. oleracea*, this mineral is responsible for the transportation of oxygen from lungs to different body parts and for the development of tissues and hormones. In addition, iron helps to have better performance during exercise and helps to increase biological functions such as reasoning, memory, and learning, and it is essential for DNA synthesis, energy, and metabolism (Tan et al., 2021). Another element found in high amounts in *P. oleracea* is magnesium, this element is essential in metabolism since it participates during energy supply processes and is necessary for activation of more than 300 enzymes, regulates muscle function, blood sugar levels, and the nervous system (Thomas et al., 2000; Ramírez de León et al., 2020).

Bioactive Compounds

Flavonoids

Flavonoids are secondary metabolites responsible for plant color, smell, and taste. They possess antibacterial, antioxidant, anti-inflammatory, and immunomodulatory properties (Liskova et al., 2021). Purslane contains apigenin-4-OaL-rhamnopyranoside, which is shown to have a hypoglycemic effect. Quercetin, which exhibits two effects on protein digestion and absorption such as suppressing protein digestion by inhibiting trypsin in the intestinal fluid, as well as promoting intestinal absorption of oligopeptides in intestinal cells (Cheng et al., 2021). Hesperidin; a flavonoid that has antioxidant, anticancer, antimicrobial properties and provides protection of cells and tissues from factors such as radiation (Iranshahi et al., 2015). Myricetin has a broad spectrum of pharmacological effects, protective functions against epoxiconazole (pesticide) -induced toxicities (Hamdi et al., 2021). Apigenin has a high potential in cancer prevention and therapy (Ahmed et al., 2021). Luteolin has neuroprotective effects both *in vivo* and *in vitro* and is considered a pharmacological agent for management and treatment of age-related neurodegenerative disorders (Nabavi et al., 2015), and Portulacanone A (2- hydroxy-5,7 dimethoxy-3-benzyl-chroman-4-one), portulacanone B (2-hydroxy 5,6,7-trimethoxy-3-benzyl-chroman-4-one), portulacanone C (5,2-dihydroxy-6,7-dimethoxy-3-benzyl-chroman-4-one), portulacanone D (5,2-dihydroxy-7-methoxy-benzylidene-chroman-4-one) whose effects are not yet known. In addition to daidzein, genistein, and genistin (Chen et al., 2019) have been associated with various therapeutic benefits such as relief of menopausal symptoms, prevention of cardiovascular diseases, and their high anticancer activity (Othman et al., 2021). On the other hand, Yang et al., (2018), mention three new compounds in *P. oleracea*: oleracona C, oleracona D, and oleracona E, these flavonoids were related to different pharmacological effects such as cardiovascular diseases, dementia, cancer, among others.

Other Phenolic Compounds

Phenolic compounds have aromatic rings more or individually coupled to hydroxyl groups. They can be classified into flavonoids and non-flavonoids, they are very heterogeneous and up to 50,000 different molecules have been identified (Ávila-Román et al., 2021). They are known from the simplest as phenolic acids to more complex compounds such as tannins. These have different functions, mainly in plant defense against UV, pathogens, etc. (Alara et al., 2021). Purslane has been mentioned to have phenolic compounds with high antioxidant activity since they can eliminate free radicals such as DPPH and reduce Fe 3+ (Li et al., 2010). Likewise, Erkan, (2012), after carrying out a purification of crude extract of purslane, reports the presence of the following acids: chlorogenic, which has antioxidant, antibacterial, hepatoprotective, cardioprotective, antiviral, antihypertensive, anti-free radical activity and is a stimulator of the nervous system (Naveed et al., 2018), caffeic which has importance in gynecology since it can protect endometrial cells against oxidative stress and could be able to prevent the development of endometriosis and complications such as pain or infertility (Jamali et al., 2021), p-coumaric improves the antioxidant capacity, and inhibits the mycelial growth of *Botrytis cinerea* and *Penicillium expansum* in sweet cherry (Liu et al., 2020), ferulic also shows to have antioxidant capacity in freshly cut apple, with potential to be added to edible coatings, thus increasing product shelf life since its quality improved not reliably, delaying oxidation reactions for 7 days of storage, in addition to inhibiting growth of pathogens such as *L. monocytogenes* (Nicolau-Lapeña et al., 2021) and rosmatriinic acid that is characterized by its inhibition of lipid peroxidation and bacterial growth. Furthermore, this compound is approved for use as a natural antioxidant and preservative in the food industry (Marchev et al., 2021). In 2009, Oliveira et al. reported the identification and quantification of two phenolic compounds in purslane: 3-caffeoylquinic acid and 5-caffeoylquinic acid (chlorogenic acids), which can play crucial roles in the regulation of lipid and glucose metabolisms, and help in different disorders such as hepatic steatosis, cardiovascular diseases, diabetes, and obesity. In addition, it was observed that the *P. oleracea* leaves contain a high number of phenolic compounds compared to the stems.

Alkaloids

Alkaloids are alkaline nitrogenous organic compounds, most of them have complex cyclic structures, the nitrogen included in the ring exerts important biological activity and these compounds can be extracted from purslane (Liu et al., 2019). They have antibacterial, anti-inflammatory, antitumor, antihypertensive, and antidepressant effects (Chen et al., 2019). Xiang et al., (2005) and Zhou et al., (2015), identified different types of alkaloids in purslane and managed to elucidate their chemical structures, from 5 oleraceins (A, B, C, D, and E) which included components already known as ρ-coumaric acid, ferulic acid and adenosine, two betalain alkaloid pigments that have attracted attention for their potential as colorants, free radical scavengers, which can help to defend against oxidative stress. In addition, reddish betacyanins and yellow betaxanthins, which are responsible for the pigmentation of stems, leaves, and flowers (Mulry et al., 2015).

Terpenes

The structure of these chemical substances is based on isopentyl units of five carbons: isopropene or 2-methyl-1,3-butadiene (Kopaczyk et al., 2020). Also known as terpenoids, they are of great importance in the pharmaceutical, cosmetic, food, and agricultural industries (Zhou and Pichersky., 2020). Using chromatography of a *P. oleracea* extract obtained with chloroform, a new clerodene diterpene portulene was identified which has antioxidant activity, in addition, other already known compounds were isolated: lupeol (triterpenoid) which has anticancer, antioxidant, anti-inflammatory, and antimicrobial properties (Liu et al. ., 2021), β-sitosterol is characterized by having anxiolytic and sedative, analgesic, immunomodulatory, antimicrobial, anticancer, anti-incencanic, anti-inflammatory, lipid-lowering effect, hepatoprotective, protective effect against NAFLD, and respiratory diseases, wound healing, antioxidants and antidiabetic effects (Babu and Jayaraman, 2020) and daucosterol (sterol) that have effects on cancer suppression, in inciting the proliferation of natural stem cells and induction of immune response (Jang et al., 2019), their structures were established by mass spectrophotometric analysis. Treatment with 70% alcohol extract from *P. oleracea* to mice with liver lesions, significantly restored liver marker enzymes and total bilirubin to almost normal values, therefore, it was shown to have hepatoprotective activity, as well as antibacterial and antifungal activity in *S. aureus subtilis, S. faecalis, E. coli, N. gonorrhea, P. aeruginosa* and *C. albicans* (Elkhayat et al., 2008).

Fatty Acids

Fatty acids are based on lipid molecules and are one of the main components of cell membranes, these acids accumulate in the form of energy reserves, are transported, and are metabolized for different purposes (Gonçalves et al., 2021). The most important fatty acids are linolenic (omega-3) and linoleic acid (omega-6), which belong to the classification of polyunsaturated fatty acids and cannot be synthesized by the human body (Zhang et al., 2021). Purslane is characterized by being rich in omega-3 fatty acids, especially α-linolenic, which is a carboxylic acid with 18 carbons and three cis double bonds, and this acid has anti-cancer, neuroprotective, anti-osteoporotic, anti-inflammatory, and antioxidant activities, they help for the good functioning of the cardiovascular system, therefore linolenic acid has multiple benefits as a nutraceutical and pharmaceutical ingredient, in addition to being safe for human consumption (Azuka et al., 2014; Kim et al., 2014). Linoleic acid constitutes the predominant proportion of polyunsaturated fatty acids in the diet, positive effects of this acid have been mentioned since it reduces the risk of cardiovascular damage, mainly coronary artery diseases (Marangoni et al., 2019). On the other hand, the presence of other essential fatty acids has been confirmed in purslane tissues such as; palmitoleic (20.96%) or omega 7, which is monounsaturated acid and has a favorable effect to reduce effects of skin and mucosa diseases and helps for skin regeneration and nutrition (Sánchez and Segundo, 2007). Palmitic acid (17.40%) is saturated and is an essential compound of the membrane, lipids secretory and transport, and signal molecules (Innis, 2016), linoleic (16.82%), and oleic (5.89%) reduce oxidative stress, preserve hepatobiliary function and support host immune and inflammatory responses (Pontes-Arruda, 2009) and stearic acids (3.46%) (Petropoulos et al., 2016; Gonella et al., 2010). Petropoulos et al., (2020), evaluated the nutritional value, chemical composition, and bioactive properties of purslane seeds, seed oils, and seed cakes and concluded that *P. oleracea* seeds are an excellent source of fatty acids such as omega-3 and omega-6.

Antioxidant Capacity

Reactive oxygen species (ROS) originate as secondary products of various metabolic pathways, likewise, they can be generated in processes of pathological inflammation, oxidases, cytokines, peroxyomas, xanthocin oxidase, NADPH oxidase, among others (Maya-Cano et al., 2021). The cause of oxidative stress is due to an imbalance between oxidants and antioxidants, favoring oxidants, and leading to disruption in oxidation-reduction signals and molecular damage (Sies, 2015). To mitigate ROS excess, there is a balance between the production and degradation of these species. Cellular mechanisms such as membranes permeability to reactive species and regulation of antioxidant enzymes allow reactive oxygen species to remain in low amounts in cells (Lushchak, 2011). Zhu et al., (2007), proved that crude polysaccharides belonging to purslane can eliminate free hydroxyl radicals showing a quantity-effect relationship, since when the concentration of crude polysaccharides in this plant was 3.5 mg/ml, the clearance rate was 50%, on the other hand, when mass concentration was 17.6 mg/ml, the clearance rate was 90%. Purslane is considered a very good source of vitamins, and antioxidants such as α-tocopherol, which has been defined as a radical chain destroyer, it is characterized by being hydrophobic, so it is kept in a lipid context, it shows antioxidant effects that are directly limited in membranes and lipoprotein domains (Azzi., 2007), ascorbic acid, also known as vitamin C, is an antioxidant that can donate electrons that allow it to be a free radical attacker to reduce oxidation states (Smirnoff, 2018) and β-carotene, which is the most common natural pigment found in nature, it can be converted into vitamin A within the human body, and has anticancer and anti- cardiovascular activities (Wang et al., 2021), including glutathione (James, 1997). Arruda et al., (2004) showed that consumption of purslane leaves has a protective effect against oxidative stress caused by vitamin A deficiency, in the heart and liver of mice treated with different diets: vitamin A deficient (DD) AIN-93G, DD supplemented with pure β-carotene (β-D) and DD supplemented with taro leaves (*Xanthosoma sagit tifolium*) (MD) or purslane (*Portulaca oleracea*) (PD) as the only source of vitamin A.

Antimicrobial Activity

Purslane flavonoids showed a bacteriostatic effect on microorganisms such as *Escherichia coli,* and *Staphylococcus aureus*, destroying bacterial cell membranes and causing spills (Chen et al., 2015). On the other hand, Saratale et al., (2020) measured spectrophotometrically the leakage of bacterial cellular components of *E. coli* and *S. aureus*, after exposing the bacteria to silver nanoparticles synthesized with *Portulaca oleracea* leaves (PNL-AgNPs) with a concentration of 25 and 35 μg/mL. Aqueous and ethereal extracts of purslane are also known to show activity against Gram-negative bacteria. The antifungal activity of purslane (50, 100, and 200 mg/ml for 150 min) has been reported against *A. niger* ATCC16404, *A. niger* IFO6661, *A. fumigatus* M5055, *A. fumigatus* Kuboyama, *C. Albicans* KULM83-0300, *C. albicans* ATCC10231, *T. mentagrophytes* EV861, and *T. mentagrophytes* HIC8185 were evaluated using a single-cell bioassay system (Oh et al., 2000). Banerjee and Mukherjee., (2002) carried out a test on the toxicity of extracts obtained with organic and aqueous solvents such as hexane, ethanol and chloroform and plant material (*Portulaca oleracea*) against *Aspergillus niger, Rhizopus artocarpi,* and *Fusarium* spp, results showed that hexane and aqueous extracts were effective against *Fusarium* spp, on the other hand, ethanol and chloroform extracts inhibited *Rhizopus artocarpi*, and finally, *Aspergillus niger* was not affected by the treatments.

Anti-Inflammatory Activity

Inflammation is an important basic pathophysiological mechanism since if there is not a sufficient inflammatory response, it can be immunodeficiency and lead from infection to cancer (Gómez-Estrada et al., 2011). Among the compounds that Hu et al., (2021) isolated from a water decoction of *Portulaca oleracea*, methyl cis-3- (3-nitro-4-hydroxyphenyl) -acrylate showed anti-inflammatory activity, which is measured as inhibition of nitric oxide generation in cells belonging to the macrophage RAW264.7. Guzmán-Heras et al., (2017) conducted a test in mice using naproxen sodium as a reference, and an extract from *Portulaca oleracea* and this extract showed inflammation decrease of the right paw in the group of experimental mice and concluded that purslane ethanolic extract contains metabolites of therapeutic interest such as alkaloids, phenols, and flavonoids, to these compounds are attributed the property of inhibiting elements that participate in the inflammation mechanisms, either for their inhibition or to prevent it from being produced with too much intensity, based on this, *P. oleracea* is considered an excellent source for obtaining natural anti-inflammatory compounds.

Hypoglycemic Activity

Portulaca oleracea seeds have hypoglycemic effect, this activity was effective and safe for patients with type 2 diabetes and has the potential to be used as adjuvant therapy since, in addition, seeds showed hypolipidaemic, and insulin resistance-reducing effects, because of the fatty acids: palmitic acid (9.2%), stearic acid (5.5%), oleic acid (18%), polysaccharides that contribute to the development of organic structures, allowing to store energy and act as a defense mechanism against some phenomena (El-Sayed, 2011). Also, Lee et al., (2020) indicated that supplementation with purslane extract improves activation of PI3K/Akt and AMPK signals in the skeletal muscle of mice, clearly noting the decrease in insulin resistance and prevention of hyperinsulinemia. In another study, it has been suggested that purslane extract increases the mass of β cells, which is of vital importance because its main function is to secrete antigen-specific antibodies and provide long-term immunity against unrecognized pathogens, in addition, they play an important role in the treatment of cancers, infections and autoimmune diseases (Stoycheva et al., 2021) and improve glucose metabolism (Ramadan et al., 2017). Finally, in another study, the aqueous extract of purslane was used during diabetic vascular complications, the treatment was carried out for 10 weeks and a decrease in blood glucose was observed, as well as in plasma triglycerides, blood pressure, LDL and HDL cholesterol levels, and insulin in diabetic mice, this represents an important advance since LDL cholesterol is the one that must be kept low, so as not to allow accumulation of cholesterol in the arteries, which leads to maintaining good blood pressure and level and/or stable health (Naeem and Khan, 2013).

Applications of Purslane (Portulaca oleracea L.) in the Food Industry

Food Preservation

Portulaca oleracea L. extract had a preservation effect on refrigerated pork for 9 days of storage, high antioxidant activity, and showed antibacterial activity against *Pseudomona aureginosa, Bacillus subtilis,* and *Bacillus cereus*. Purslane treatments significantly decreased microbial growth, lipid oxidation was delayed, and superoxide dismutase (SOD) enzymes responsible for catalyzing the detoxification of the superoxide anion were increased, decreasing and avoiding the formation of reactive oxygen species (ROS) cascade is considered among the best antioxidants that help to reduce cellular aging and glutathione peroxidase (GHS-Px), maintaining protective effects on some neurodegenerative disorders (Parkinson's disease, Alzheimer's, cerebral ischemia, and compulsive disorders), finally, treatments with 1% and 0.50% of the extract maintained visual quality compared to the control treatment after nine days (Miller., 2012; Sharma., 2021; Fan et al., 2019). In other research, purslane extract also affected inhibition of browning of freshly sliced potatoes, the effect was observed in storage conditions for eight days at 4°C, therefore this alternative is economical, with high content of polyphenols and alkaloids, and can be considered as an excellent anti-darkening strategy to maintain vegetable quality (Liu et al., 2019).

Fermentation of *Portulaca oleracea* L.

An experiment was carried out to explore purslane as a potential substrate to develop new functional products through lactic acid fermentation, where *Lactobacillus kunkeii* B7 and *Lactobacillus plantarum* POM1 were used, results showed an approximate 30% decrease in oxalic acid. The conversion of oleraceins by *L. kunkeii* B7 was also observed, which is possibly a cause of the increase in antioxidant activity. Finally, it was concluded that fermentation with lactic acid bacteria of *Portulaca oleracea* is an alternative with great potential for the development of functional foods (Filannino et al., 2017).

Foods Fortified with *Portulaca oleracea* L.

Mastud et al., (2018) reported a trial in which purslane powder was used for the elaboration of cookies and ice cream, the powder was obtained by drying through an oven at 60°C, an evaluation was made for its chemical analysis, and later 1, 2 and 3 g were used for ice cream, and 2, 4, 6 and 8 g for cookies, results showed that purslane powder contains high amounts of proteins and antioxidants, in addition to this, achieved sensory acceptance from the public with the values of 2 g for ice cream and 6 g for cookies.

On another occasion, Hussien and Salem. (2016) fortified rice flour with three levels of purslane powder (5%, 10%, and 15%) and then subjected to a physical-chemical and sensory analysis. The addition of purslane to the flour, significantly increased the protein, ash, and fiber content of the prepared snacks, in addition, there was an increase in the total phenol content, which was proportional to the amount of purslane added (179.89 mg/100 g). Therefore, adding functional ingredients to food matrices can provide excellent results, not only for making satisfying or satisfying foods but also for adding nutrients to the diet. On the other hand, Melilli et al., (2020) evaluated incorporation of 0%, 5%, 10%, and 15% of dry purslane flour on the rheological, sensory and nutritional quality of wheat bread, obtaining that with the incorporation of 5%, better rheological characteristics and good antioxidant properties were obtained, as well as a high amount of fatty acids, since in 10% and 15% there was no increase in linoleic acid (omega-3) and α-linolenic acid (omega-3). 6), this can be attributed to cooking.

Purslane Drinks

According to Wang et al. (2016), a purslane drink can easily be prepared after cleaning, washing, and steaming it. The drink has excellent antioxidant activity and DPPH clearance due to the extraction rate of active compounds. It is a tea that contains 50% pumpkin flower, 20% *Grosvenor momordica* flowers, and 30% purslane. The last ingredient can reduce fatigue, activate the body, prevent cardiovascular diseases, among other benefits (Chen et al., 2019). In recent years, interest in purslane and its benefits has increased around the world (Table 2).

Table 2. Patented applications of *P. oleracea* in foods

Tittle	Inventor	Description
Portulaca oleracea L. extract feed additive and preparation method thereof	Jiaming Chen, Xuefeng Chen, Ling Lu, Ting Shu, Zili Chen, Danrong Zheng, Shunjie Huang, Ziqiang Chen.	An additive created for animals, which is capable of improving the animal survival rate, reducing diseases incidence and stress, and increasing animal productivity.
Chewable purslane tablet	Xie Xiaofeng	*Portulaca oleracea* tablet with therapeutic efficiency, health care, and disease prevention.
Portulaca oleracea drink containing antimicrobial and antitumor effects	Bae Ji Hyun, Bae Hyo Jin	Functional drink with antimicrobial, anticancer effect, even elimination of nicotine.
Portulaca oleracea edible oil	Li, Huamin	Edible oil to prevent and treat arteriosclerosis, intestinal diseases, cerebral infarction.
Portulaca oleracea tea bags	Yougen	Pleasant taste, decrease in blood fat, asteclerosis, increases immunity, anticancer, prevention of obesity.

Biosafety

Currently, here is a lack of information on the negative effects or possible problems that purslane can cause, however, a risk factor is the high levels of oxalate in the plant (Moreau and Savage, 2009). Very high oxalate values have been found throughout the plant, and even in purslane leaves, values from 2.55 to 12.94 of total oxalate in kg-1 of fresh weight were reported (Mishra et al., 2020). Poeydomenge and Savage, (2007), reported that purslane leaves, stems, and buds contained 235, 56, and 91 mg of total oxalate per 100 g in fresh weight, respectively. Oxalate has no use after absorption, it is only transported to kidneys and expelled in urine as waste, although, it is a risk factor since calcium oxalate crystals can develop, which are what produce the well-known kidney stones.

Future Trends

Reviewing in a general way the recent advances on purslane properties, there have been developments in the food industry, for example, the development of teas, chewable tablets, drinks, among others. But, it is disturbing that there is relevant information about this plant, however, it is not very well known, one way to be able to investigate further is to characterize purslane molecularly using genes from other plants, to later sequence them and learn more about this plant species, genus, and family, even carry out fermentation with lactic acid bacteria and take advantage of it, to develop a functional food.

Conclusion

This document presents an updated review on taxonomy, applications, phytochemicals, phenolic compounds of purslane (*Portulaca oleracea* L.). Different investigations mention diverse biological activities of this plant, including anti-inflammatory, antimicrobial, hepatoprotective, hypolipominant, antioxidant, etc. Based on the available literature, it can be concluded that purslane has great prophylactic and therapeutic potential in animals, and it also has applications in the food industry for the production and preservation of food, even for the development of functional ingredients. There is still much to know, from its exact origin, as well as its use in recent technologies, since there is only information on *P. oleracea* extracts, and information is lacking on topics such as fermentation of its fiber and determining its synergy with lactic acid bacteria to increase its functionality.

Acknowledgments

This study had financial support from the Universidad Autonoma de Coahuila. CGR received financial support from the National Council of Science and Technology-Mexico (CONACyT) for her postgraduate studies.

Conflict of Interest

The authors declare no conflict of interest.

References

Ahmed, S. A., Parama, D., Daimari, E., Girisa, S., Banik, K., Harsha, C., Dutta, U., Kunnumakkara, A. B. (2021). Rationalizing the therapeutic potential of apigenin against cancer. *Life Sciences,* 267.

Alara, O. R., Abdurahman, N. H., Ukaegbu, C. I. (2021). Extraction of phenolic compounds: A review. *Current Research in Food Science,* 4, 200-214.

Arruda, S. F., Siqueira, E. M. A., Souza, E. M. T. (2004). Malanga (*Xanthosoma sagittifolium*) and Purslane (*Portulaca oleracea*) leaves reduce oxidative stress in vitamin A-deficient rats. *Annals of Nutrition and Metabolism,* 48(4), 288-295.

Ávila-Román, J., Soliz-Rueda, J. R., Bravo, F. I., Aragonés, G., Suárez, M., Arola-Arnal, A., Mulero, M., Salvadó, M. J., Arola, L., Torres-Fuentes, C., Muguerza, B. (2021). Phenolic compounds and biological rhythms: who takes the lead? *Trends in Food Science & Technology,* 113, 77-85.

Azuka, O. I., Ayalokunrin, M. B., Abu, O. L. (2014). A review on Portulaca oleraceae (purslane) plant- its nature and biomedical benefits. *International Journal of Biomedical Research,* 5, 75-80.

Azzi, A. (2007). Molecular mechanism for α-tocopherol action. *Free Radical Biology and Medicine,* 43(1), 16-21.

Babu, S., Jayaraman, S. (2020). An update on β-sitosterol: a potential herbal nutraceutical for diabetic management. *Biomedicine & Pharmacotherapy,* 131.

Banerjee, G., Mukherjee, A. (2002). Biological activity of a common weed: *Portulaca oleracea* L-II. Antifungal activity. *Acta Botánica Hungárica,* 44(3), 205-208.

Berrones-Morales, M., Torres-Acosta, R. I., Rangel-Torres, E., Padrón-Torres, E. (2020). Siembra y cosecha de la verdolaga. *Guía técnica del cultivo de la verdolaga,* 21-26. [Sowing and harvesting of purslane. *Purslane cultivation technical guide*]

Bird, R. P., Eskin, N. A. M. (2021). The emerging role of phosphorus in human health. *Advances in Food and Nutrition Research.*

Calderón de Rzedowski, G. (2001). "Portulaceae." En: Rzedowski, G. C. de, J. Rzedowski y colaboradores. *Flora fanerogámica del Valle de México.* 2ª. Ed., Instituto de Ecología, A. C. y Comisión Nacional para el Conocimiento y Uso de la Biodiversidad. Pátzcuaro (Michoacán), 1406. ["Portulaceae." In: Rzedowski, G. C. de, J. Rzedowski et al. *Phanerogamic flora of the Valley of Mexico.* 2nd. Ed., Institute of Ecology, A. C. and National Commission for the Knowledge and Use of Biodiversity. Patzcuaro (Michoacán), 1406]

Chen, G. N., Sun, F. L., Yan, Y. R., Liu, M. M. (2015). Study on antibacterial mechanism of flavonoids from purslane. *Chemistry & Bioengineering,* 32(10), 34-37.

Chen, D., Yao, J., Liu, T., Zhang, H., Li, R., Zhang, Z., Gu, X. (2019). Research and application of *Portulaca oleraceae* in pharmaceutical area. *Chinese Herbal Medicines,* 11, 150-159.

Cheng, Y., Liu, Y., Chen, D., Zhou, Y., Yu, S., Lin, H., Kai-Liao, C., Lin, H., Xu, P., Huang, M. (2021). Dual effects of quercetin on protein digestion and absorption in the digestive tract. *Food Chemistry,* 358.

De Lorenzo-Cáceres, J. M. S. (2021). *Aportación al conocimiento del género Portulaca L. (Portulacaceae) en España.* Consultado en mayo de 2021. [*Contribution to the knowledge of the genus Portulaca L. (Portulacaceae) in Spain*]

Desta, M., Molla, A., Yusuf, Z. (2020). Characterization of physico-chemical properties and antioxidant activity of oil from seed, leaf and stem of purslane (*Portulaca oleracea* L). *Biotechnology Reports,* 27.

Egea-Gilabert, C., Ruiz-Hernández, M. V., Parra, M. A., Fernández, J. A. (2014). Characterization of purslane (*Portulaca oleracea* L.) accessions: Suitability as ready-to-eat producto. *Scientia Horticulturae,* 172, 73-81.

Elcinto, M. A. (2000). El potasio para la salud. *Medicina naturista,* 1, 17-19.

Elkhayat, E. S., Ibrahim, S. R. M., Aziz, M. A. (2008). Portulene, a new diterpene from *Portulaca oleracea* L. *Journal of Asian Natural Products Research,* 10(11), 1039-1043.

El-Sayed, M. I. (2011). Effects of *Portulaca oleracea* L. seeds in treatment of type-2 diabetes mellitus patients as adjunctive and alternative therapy. *Journal of Ethnopharmacology,* 137(1), 643-651.

El-Shora, H. M., Abd El-Gawad, A. M. (2015). Physiological and biochemical responses of Curcubita pepo L. mediated by *Portulaca oleracea* L. allelopathy. *Fresenius Environmental Bulletin,* 24(1B), 386-393.

Erkan, N. (2012). Antioxidant activity and phenolic compounds of fractions from Portulaca oleracea L. *Food Chemistry*, 133 (3), 775-781.

Etnobotánica. (2021). "La verdolaga, tradición, salud, alimentación y cosmética natural. Base de datos en línea." La verdolaga, tradición, salud, alimentación y cosmética natural Etnobotánica (etnobotanica.net). Consultado en mayo del 2021. ["Purslane, tradition, health, food and natural cosmetics. Online database." *Purslane, tradition, health, food and natural cosmetics Ethnobotany (etnobotanica.net).*]

Fan, X. J., Liu, S. Z., Li, H. H., He, J., Feng, J. T., Zhang, X., Yan, H. (2019). Effects of Portulaca oleracea L. extract on lipid oxidation and color of pork meat during refrigerated storage. *Meat Science,* 147, 82-90.

Filannino, P., Di Cagno, R., Trani, A., Cantatore, V., Gambacorta, G., Gobbetti, M. (2017). Lactic acid fermentation enriches the profile of biogenic compounds and enhances the functional features of common purslane (*Portulaca oleracea* L.) *Journal of Functional Foods,* 39, 175-185.

Fontana, E., Hoeberechts, J., Nicola, S., Cros, V., Palmegiano, G. B., Peiretti, P. G. (2006). Nitrogen concentration and nitrate/ammonium ratio affect yield and change the oxalic acid concentration and fatty acid profile of purslane (*Portulaca olearacea* L.) grown in a soilless culture system. *Journal of the Science of Food and Agriculture,* 86, 2417-2424.

Gómez-Estrada, H. A., González-Ruiz, K. N., Domingo-Medina, J. (2011). Actividad antiinflamatoria de productos naturales. *Boletín Latinoamericano y del Caribe de Plantas Medicinales y Aromáticas,* 10(3), 182-217.

Gonçalves, A. M. M., Rocha, P. C., Marques, J. C., Gonçalves, F. J. M. (2021). Fatty acids as suitable biomakers to assess pesticide impacts in freshwater biological scales- A review. *Ecological Indicators,* 122.

Gonnella, M., Charfeddine, M., Conversa, G., Santamaría, P. (2010). Purslane: a review of its potential for health and agricultural aspects. *Plant Science Biotechnology,* 4, 131-136.

Guzmán-Heras, L. E., García-Mir, V., Cuesta-Rubio, O, Jaramillo-Jaramillo, C. G., Ramón-Japón, G. E. (2017). Chemical composition and antiinflammatory activity of the extract from aerial parts of *Portulaca oleracea* (purslane). *Revista Cubana de Farmacia,* 51(1). [*Cuban Journal of Pharmacy*]

Hambi, H., Abid-Essefi, S., Eyer, J. (2021). Neuroprotective effects of myricetin on epoxiconazole-induced toxicity in F98 cells. *Free Radical Biology and Medicine,* 164, 154-163.

Hu, S., Chai, W. C., Xu, L., Li, S., Jin, C., Zhu, R., Yang, L., Zhang, R., Tang, K., Li, P., Yang, E., Chang, W., Shen, T., Semple, S., Venter, H., Xiang, L. (2021). Catecholic alkaloid sulfonates and aromatic nitro compounds from Portulaca oleracea and screening of their anti-inflammatory and anti-microbial activities. *Phytochemistry,* 181.

Hussien, H. A., Salem, E. M. (2016). Development of gluten free snacks fortified with purslane (Portulaca oleracea) powder. *Journal of Food and Nutrition Sciences,* 4(6), 136-144.

Innis, S. M. (2016). Palmitic acid in early human development. *Critical Reviews in Food Science and Nutrition,* 56(12), 1952-1959.

Iranshahy, M., Rezaee, R., Parhiz, H., Roohbakhsh, A., Soltani, F. (2015). Protective effects of flavonoids against microbes and toxins: The cases of hesperidin and hesperetin. *Life Sciences,* 137, 125-132.

Iranshahy, M., Javadi, B., Iranshani, M., Jahanbakhsh, S. P., Mahyari, S., Hassani, F. V., Karimi, G. (2017). A review of traditional uses, phytochemistry and pharmacology of *Portulaca oleracea* L. *Journal of Ethnopharmacology,* 205, 158-172.

Jamali, N., Mostafavi-Pour, Z., Zal, F., Kasraeian, M., Poordast, T., Nejabat, N. (2021). Antioxidant ameliorative effect of caffeic acid on the ectopic endometrial cells separated from patients with endometriosis. *Taiwanese Journal of Obstetrics and Gynecology,* 60(2), 216-220.

James, A. D. (1997). The Green Pharmacy: New discoveries in herbal remedies for common diseases and conditions from the world's foremost authority on healing herbs. Published by Peggy K. Duke, New York, USA.

Jang, J., Kim, S. M., Yee, S. M., Kim, E. M., Lee, E. H., Choi, H. R., Lee, Y. S., Yang, W. K., Kim, H. Y., Kim, K. H., Kang, H. S., Kim, S. H. (2019). Daucosterol suppresses dextran sulfate sodium (DSS)- induced colitis in mice. *International Immunopharmacology,* 72, 124-130.

Kim, K. B., Nam, Y. A., Kim, H. S., Hayes, A. W., Lee, B. M. (2014). α-Linolenic acid: Nutraceutical, pharmacological and toxicological evaluation. *Food and Chemical Toxicology,* 70, 163-178.

Kopaczyk, J. M., Wargula, J., Jelonek, T. (2020). The variability of terpenes in conifers under developmental and environmental stimuli. *Environmental and Experimental Botany,* 180.

Lara, L. J., Eguea-Gilbert, C., Niñirola, D., Conesa, E., Fernández, J. A. (2011). Effect of aeration of the nutrient solution on the growth and quality of purslane (Portulaca oleracea). *Journal of Horticultural Science and Biotechnology,* 86, 603-610.

Lee, J. H., Park, J. E., Han, J. S. (2020). Portulaca oleracea L. extract reduces hyperglycemia via PI3/Akt and AMPK pathways in the skeletal muscles of C57BL/Ksj-db/db mice. *Journal of Ethnopharmacology,* 260.

Li, D. F., Jia, D. Y., Du, X., Yao, K. (2010). Antioxidant activity of phenolic extracts from Portulaca oleraceae L. *China Oils and Fats,* 35 (12), 41-43.

Liskova, A., Samec, M., Koklesova, L., Samuel, S. M., Zhai, K., Al-Ishaq, R. K., Abotaleb, M., Nosal, V., Kajo, K., Ashrafizadeh, M., Zarrabi, A., Brockmueller, A., Shakibaei, M., Sabaka, P., Mozos, I., Ullrich, D., Prosecky, R., La Rocca, G. (2021). Flavonoids against the SARS-CoV-2 induced inflammatory storm. *Biomedicine & Pharmacotherapy,* 138.

Liu, C., Yang, S., Wang, K., Bao, X., Liu, Y., Zhou, S., Liu, H., Qiu, Y., Wang, T., Yu, H. (2019). Alkaloids from traditional chinese medicine against hepatocellular carcinoma. *Biomedicine & Pharmacotherapy,* 120.

Liu, X., Yang, Q., Lu, Y., Li, Y., Li, T., Zhou, B., Qiao, L. (2019). Effect of purslane (*Portulaca oleracea* L.) extract on anti-browning of fresh-cut potato slices during storage. *Food Chemistry,* 283, 445-453.

Liu, X., Ji, D., Cui, X., Zhang, Z., Li, B., Xu, Y., Chen, T., Tian, S. (2020). ρ-Coumaric acid induces antioxidant capacity and defense responses of sweet cherry fruit to fungal pathogens. *Postharvest Biology and Technology,* 169.

Liu, K., Zhang, X., Xie, L., Deng, M., Chen, H., Song, J., Long, J., Li, X, Luo, J. (2021). Lupeol and its derivates as anticancer and anti-inflammatory agents: molecular mechanisms and therapeutic efficacy. *Pharmacological Research,* 164.

Lushchak, V. I. (2011). Adaptative response to oxidative stress: Bacteria, fungi, plants and animals. *Comparative Biochemistry and Physiology Part C: Toxicology & Pharmacology,* 153 (2), 175-190.

Marangoni, F., Agostoni, C., Borghi, C., Catapano, A. L., Cena, H., Ghiselli, A., Poli, A. (2019). Dietary linoleic acid and human health: Focus on cardiovascular and cardiometabolic effects. *Atherosclerosis,* 292, 90-98.

Marchev, A. S., Vasileva, L. V., Amirova, K. M., Savova, M. S., Koycheva, I. K., Balcheva-Sivenova, Z. P., Vasileva, S. M., Georgiev, M. I. (2021). Rosmarinic acid - From bench to valuable applications in food industry. *Trends in Food Science & Technology.*

Martínez-López, J. N., Rodríguez-Castillejos, G. C., Martínez-Ávalos, J. G., Mora-Olivo, A. (2020). Generalidades de la verdolaga. *Guía técnica del cultivo de la verdolaga,* 11-16. [Purslane overview. *Purslane cultivation technical guide*]

Mastud, S. K., Mote, G. V., Sahoo, A. K. (2018). Development of value added products by using purslane (*Portulaca oleracea*). *Journal of Pharmacognosy and Phytochemistry,* 7(4), 1761-1766.

Maya-Cano, D. A., Arango-Varela, S., Santa-González, G. A. (2021). Phenolic compounds of blueberries (*Vaccinium* spp) as a protective strategy against skin cell damage induced by ROS: A review of antioxidant potential and antiproliferative capacity. *Heliyon,* 7 (2).

Melilli, M. G., Di Stefano, V., Sciacca, F., Pagliaro, A., Bognanni, R., Scandurra, S., Virzi, N., Gentile, C., Palumbo, M. (2020). Improvement of fatty acid profile in durum wheat breads supplemented with *Portulaca oleracea* L. quality traits of purslane-fortified bread. *Foods,* 9(6), 764.

Mera-Ovando, L. M., Castro-Lara, D., Bye-Bottler, R. (2010). *Importancia de la verdolaga en México*. 1-10.

Miller, A. F. (2012). Superoxide dismutases: ancient enzymes and new insights. *FEBS Letters,* 586(5), 585-595.

Mishra, V., Chugh, V., Dwivedi, S. V., Sharma, K. D. (2020). Food and nutraceuticals value of purslane (*Portulaca oleracea* L.): An overview. *The Pharma Innovation Journal,* 9(7), 419-424.

Moreau, A. G., Savage, G. P. (2009). Oxalate content of purslane leaves and the effect of combining them with yoghurt or coconut products. *Journal of Food Composition and Analysis,* 22(4), 303-306.

Mulry, K. R., Hanson, B. A., Dudle, D. A. (2015). Alternative strategies in response to saline stress in two varieties of *Portulaca oleraceae* (purslane). *PLoS One,* 10.

Nabavi, S. F., Braidy, N., Gortzi, O., Sobarzo-Sanchez, E., Daglia, M., Skalicka-Wozniak, K., Nabavi, S. M. (2015). Luteolin as an anti-inflammatory and neuroprotective agent: A brief review. *Brain Research Bulletin,* 119(A), 1-11.

Naeem, F., Khan, S. H. (2013). Purslane (*Portulaca oleracea* L.) as phytogenic substance- a review. *Journal of Herbs, Spices & Medicinal Plants,* 19(3), 216-232.

Naveed, M., Hejazi, V., Abbas, M., Kamboh, A. A., Khan, G. J., Shumzaid, M., Ahmad, F., Babazadeh, D., FangFang, X., Modarresi-Ghazani, F., WenHua, L., XiaoHui, Z. (2018). Chlorogenic acid (CGA): A pharmacological review and call for further research. *Biomedicine & Pharmacotherapy,* 97, 67-74.

Nemzer, B., Al-Taher, F., Abshiru, N. (2020). Phytochemical composition and nutritional value of different plant parts in two cultivated and wild purslane (*Portulaca oleracea* L). genotypes. *Food Chemistry,* 320.

Nicolau-Lapeña, I., Aguiló-Aguayo, I., Kramer, B., Abadías, M., Viñas, I., Muranyi, P. (2021). Combination of ferulic acid with Aloe vera gel or alginate coatings for shelf-life prolongation of fresh-cut apples. *Food Packaging and Shelf Life,* 27.

Ocampo, G., Columbus, J. T. (2012). Molecular phylogenetics, historical biogeography, and chromosome number evolution of Portulaca (*Portulacaceae*). *Molecular Phylogenetics and Evolution,* 63(1), 97-112.

Oh, K. B., Chang, I. M., Hwang, K. J., Mar, W. (2000). Detection of antifungal activity in Portulaca oleracea by a single-cell bioassay system. *Phytother Res,* 14(5), 329-3.

Oliveira, I., Valentao, P., Lopes, R., Andrade, P. B., Bento, A., Pereira, J. A. (2009). Phytochemical characterization and radical scavenging activity of *Portulaca oleraceae* L. leaves and stems. *Microchemical Journal,* 92 (2), 129-134.

Othman, M. M. A., Sehata, M. M. E., Elnaggar, S. R. Y. (2021). Multifaceted implementation of nanotechnology in ameliorating therapeutic efficacy of soy phytoestrogens: Comprehensive review on the state of art. *Journal of Drug Delivery Science and Technology,* 61.

Petropoulos, S., Karkanis, A., Martins, N., Ferreira, I. C. F. R. (2016). Phytochemical composition and bioactive compounds of common purslane

(*Portulaca oleracea* L.) as affected by crop management practices. *Trends in Food Science & Technology,* 55, 1-10.

Petropoulos, S. A., Fernandes, A., Arampatzis, D. A., Tsiropoulos, N. G., Petrovic, J., Sokovic, M., Barros, L., Ferreira, I. C. F. R. (2020). Seed oil and seed oil byproducts of common purslane (*Portulaca oleraceae* L.): A new insight to plant-based sources rich in omega-3 fatty acids. *LWT- Food Science and Technology,* 123.

Poeydomenge, G. Y., Savage, G. P. (2007). Oxalate content of raw and cooked purslane. *Journal of Food, Agriculture and Environment,* 5, 124-128.

Pontes-Arruda, A. (2009). Biological benefits of an oleic acid-rich lipid emulsion for parenteral nutrition. *Clinical Nutrition Supplements,* 4(1), 19-23.

Pu, F., Chen, N., Xue, S. (2016). Calcium intake, calcium homeostasis and health. *Food Science and Human Wellness,* 5(1), 8-16.

Ramadan, B. K., Schaalan, M. F., Tolba, A. M. (2017). Hypoglycemic and pancreatic protective effects of Portulaca oleracea extract in alloxan induced diabetic rats. *BMC Complementary and Alternative Medicine,* 17(1), 37.

Ramírez de León, J. A., Uresti-Marín, R. M., Castañón-Rodríguez, J. F. (2020). Valor nutricional de la verdolaga. *Guía técnica del cultivo de la verdolaga,* 51-54.

Román-Cortés, N. R., García-Mateos, M. R., Castillo-González, A. M., Sahagún-Castellanos, J., Jiménez-Arellanes, M. A. (2018). Características nutricionales y nutracéuticas de hortalizas de uso ancestral en México. *Revista Fitotecnia Mexicana*, 41 (3), 245-253. [Nutritional and nutraceutical characteristics of vegetables of ancestral use in Mexico. *Mexican Fitotecnia Magazine*]

Sánchez, P., Segundo, M. J. (2007). Omega 7. *Offarm,* 26(7), 58,64.

Saratale, G. D., Saratale, R. G., Cho, S. K., Ghodake, G., Bharagava, R. N., Park, Y., Mulla, S. I., Kim, D. S., Kadam, A., Nair, S., Shin, H. S. (2020). Investigation of photocatalytic degradation of reactive textile dyes by Portulaca oleracea-functionalized silver nanocomposites and exploration of their antibacterial and antidiabetic potential. *Journal of Alloys and Compounds,* 833.

Servicio Nacional de Inspección y Certificación de Semillas. (2017). Verdolaga (*portulaca oleracea* L.) Base de datos en línea. *Verdolaga (portulaca oleracea L.)** |Servicio Nacional de Inspección y Certificación de Semillas|Gobierno|gob.mx (www.gob.mx). Consultado en mayo del 2021.

Sharma, G., Shin, E. J., Sharma, N., No, S. Y., Mai, H. N., Nguyen, B. E., Jeong, J. H., Lei, X. G., Kium, H. C. (2021). Glutathione peroxidase-1 and neuromodulation: novel potentials of an old enzyme. *Food and Chemical Toxicology,* 148.

SIAP. (2019). *Servicio de Información Agroalimentaria y Pesquera. Base de datos en línea.* Verdolaga_compressed.pdf (www.gob.mx). Consultado en mayo del 2021. [*Agrifood and Fisheries Information Service.* Online database.]

Sies, H. (2015). Oxidative stress: a concept in redox biology and medicine. *Redox Biology,* 4, 180-183.

Smirnoff, N. (2018). Ascorbic acid metabolism and functions: a comparison of plants and mammals. *Free Radical Biology and Medicine,* 122, 116-129.

Stoycheva, D., Simsek, H., Weber, W., Hauser, A. E., Klotzsch, E. (2021). External cues to drive B cell function towards immunotherapy. *Acta Biomaterialia.*

Tan, Z., Lu, P., Adewole, D., Diarra, M. S., Gong, J., Yang, C. (2021). Iron requirement in the infection of Salmonella and its relevance to poultry health. *Journal of Applied Poultry Research,* 30(1).

Thomas, S. B., Fortmann, M., Wenish, S. (2000). Magnesio, un elemento clave para la salud. *Alimentación, equipos y tecnología,* 19(1), 119-121. [Magnesium, a key element for health. *Food, equipment and technology*]

Uddin, M. K., Juraimi, A. S., Anwar, F., Hossain, M. A., Alam, M. A. (2012). Effect of salinity on proximate mineral composition of purslane (*Portulaca oleracea* L.). *Australian Journal of Crop Science,* 6(12), 1732-1736.

Verdolaga (Portulaca oleracea) (2021). www.naturalista.mx/taxa/58991-Portulaca-oleracea#Taxonomía.

Vibrans, H. (2009). *Malezas de México,* 25 abril 2021. Chenopodium ambrosioides - ficha informativa (conabio.gob.mx). [*Mexican weeds*]

Wang, H. K., Mao, H. H., Zeng, T. T., Zhang, C., Zhang, T. (2016). *The method of preparation purslane beverage.* CN106173764A.

Wang, L., Liu, Z., Jiang, H., Mao, X. (2021). Biotechnology advances in β-carotene production by microorganisms. *Trends in Food Science & Technology,* 111, 322-332.

Xiang, L., Xing, D., Wang, W., Wang, W., Wang, R., Ding, Y., Du, L. (2005). Alkaloids from Portulaca oleracea L. *Phytochemistry,* 66, 2595-2601.

Yang, X., Zhang, W., Ying, X., Stien, D. (2018). New flavonoids from Portulaca oleracea L. and their activities. *Fitoterapia,* 127, 257-262.

Zhang, R., Sun, X., Vidyarthi, S. K., Wang, F., Zhang, Y., Pan, Z. (2021). Characteristics of fatty acids in the Chinese jujube fruits (ZiZiphus jujuba mil). *Journal of Agriculture and Food Research,* 4.

Zhou, Y. X., Xin, H. L., Rahman, K., Wang, S. J., Peng, C., Zhang, H. (2015). Portulaca oleracea L: a review of phytochemistry and pharmacological effects. *Journal of Biomedicine and Biotechnology,* 5.

Zhou, F., Pichersky, E. (2020). More is better: the diversity of terpene metabolism in plants. *Current Opinion in Plant Biology,* 55, 1-10.

Zhu, X. H., Wu, X. Y., Yang, L. Q., Fan, Q. Y., Mao, G. H. (2007). Study on extraction and ability to scavenge hydroxyl radicals for Polysaccharides from *Portulaca oleracea* L. *Journal of Jiangsu University (Medicine Edition),* 17(1), 57-60.

Chapter 4

Amaranthus spp. as a High Nutritional Value Food Alternative

Guadalupe Linor Pech Damian, Adriana C. Flores Gallegos, Juan Alberto Ascacio Valdés, Sendar Daniel Nery Flores, Raúl Rodríguez Herrera and Sonia Yesenia Silva Belmares*

Department of Food Research. School of Chemistry.
Universidad Autónoma de Coahuila. Boulevard Venustiano Carranza, Saltillo, Coahuila, Mexico

Abstract

Currently, amaranths are considered alternative crops with food potential and comprise more than 350 species. They have been used in human food for more than 1000 years. However, the culinary traditions that included them daily in the diet have been lost. This genus is distributed throughout the world and is characterized by its resistance to adverse climates, and in most countries, these plants are considered invasive. However, research in recent decades has demonstrated the importance of *Amaranthus* spp. as an alternative source of nutrients (protein, fiber, minerals, and carbohydrates), which could give a favorable turn in the diversity of agroecosystems, health, cultural richness, and technological innovation in the food industry. On the other hand, biodiversity

* Corresponding Author's E-mail: yesenia_silva@uadec.edu.mx.

In: Food Processing
Editors: Myriam Huijs
ISBN: 978-1-68507-570-5

encompasses different levels of organization that include composition, structure and function to understand the importance of highly valued edible plants, such as *Amaranthus* species. From *Amaranthus* plants, stems, leaves, flowers, and seeds can be used to obtain micro-macro nutrients and phytochemicals. In addition, extracts from their roots are suitable as bioherbicides. Additionally, it is recommended to study the future impact of technological and agricultural developments in food security in developing countries.

Keywords: edible plants, *Amaranthus* spp., nutrition, food, alternatives

Introduction

Biodiversity is a concept that encompasses different levels of biological organization that includes plants, animals, fungi and microorganisms that inhabit a specific space. Additionally, it includes ecological-evolutionary processes that occur at the level of genes, species, ecosystems and landscapes (CONABIO, 2020). As well, the cultural diversity of each region shows that human beings and nature have maintained a relationship that gives rise to a clear food imbalance in different geographical areas. Therefore, the plants proposed as alternative crops stand out for their tolerance to abiotic stress. Additionally, According to International Center for Biosaline Agriculture, the conservation genetic improvement and selection of its crops are recommended to improve food security (Elouafi, 2020). Mexico presents one of the richest biocultural diversities worldwide, which is reflected in its outstanding and varied gastronomy, despite the little information generated on edible resources at the local level and the nutrition of rural and indigenous communities (Manzanero-Medina et al., 2020). Several studies on phylogenetic resources for the Food and Mexican Agriculture office revalue native crops, promote their conservation, and use them as alternative foods. Non-conventional food plants, ancestral grains, flowers, meliponicultural products, and edible insects are sources of nutrients, besides contributing to food, increasing the local economy (Barros et al., 2021). In developing countries, small farmers produce most of their food, however, they are vulnerable to poverty, limited diversity of their diets, and poor nutrition, so a change in social behavior that supports the utilization of diverse basic baskets is necessary (Nabuuma et al., 2021).

Edible Alternative Crops

The interaction between society-nature is built from the experiences transmitted among people from generation to generation. These experiences are regulated by complex belief systems, symbolism, customs, worldview, ecological knowledge, and applied management strategies, marking eating patterns at the regional and local level (Farfán-Heredia et al., 2018).

Mexico is a megadiverse country and is considered the fifth nation in floristic richness since it contains between 25,000 and 30,000 species of plants. The ethnobotanical database of Mexican useful plants (BADEPLAM) has records of 7,461 useful plants of which 2,168 are edible (Mapes and Barsuto, 2016). According to this, the traditional knowledge about the use of plants of the Mexican indigenous communities comes from their ancestors (Ramos et al., 2019). The Florentine Codex, the Badiano Manuscript and the History of the Plants of New Spain (Picó y Nuez, 2000) are some of the manuscripts that have made it possible to reconstruct the Mesoamerican pre-Hispanic horticultural framework on the knowledge of the plants of the Aztecs through the history. The manuscripts describe little-known edible crops, selected by humans through experience. Additionally, they describe different ways of managing wild plants, weeds and plants domesticated by the Nahua and Mixtecs in the Balsas river basin, Mexico. They also include the ways of managing a collection of wild populations, the tolerance of plants in clearings of natural vegetation, the improvement and protection of species. They also describe how to sow or plant propagules and transplant entire individuals of weeds and wild vegetation in controlled environments (Casas et al., 1996). Likewise, the Mesoamerican population implemented the agriculture of corn, beans, squash and chilli peppers, which are now important worldwide. However, little attention was paid to lesser-known crops that were part of their food and medicine (Picó and Nuez, 2000). Mexico's traditional agricultural spatial components, its weeds, and wild plants form a system of several relatively well-defined and ecologically distinct factors, among which cultivated fields and home gardens are distinguished. However, field margins and rural habitats are also part of the productive system, as are old fields, secondary and primary forests, and aquatic environments. In these spaces, Mesoamerican farmers planted some less wild or semi-domesticated European plants and integrated them into these systems (Vibras, 2016; Guzmán-Lucio et al., 2013). In Mexico, a relatively large number of plants were introduced and became part of its food traditions, standing out for their collection in agroecosystems, harvests in wild areas, and appearing as important crops (only some). Colloquially, some of these species are known as amaranths, this classification of plants although currently managed as native edible species, in their great majority were introduced approximately 500 years ago (Manzanero et al., 2020).

The word amaranth or quelite (in Spanish) comes from the Nahuatl word quilitl which means vegetable or tender edible plant, these plants have been named and consumed by different cultures among which, One cans mention the following native languages: ntiyia (Mazatec), yiwa, or yube (Mixtec), quilitl or quilit (Náhuatl and Náhuat), k'ani (ñahñu), xakua (Purépecha), guilibá (Rarámuri), akw'aal (Tenek), ivagi (Tepehuan), kaka (Totonaco), bokitah (Tzeltal) and itaj (Tzotzil) (Mapes and Barsuto, 2016). These species are characterized by small, soft, and flexible stems, others are tall and hard bushes from which branches, leaves, fruits, or flowers are formed (ASERCA, 2018).

Encouraging the development of these edible plants around residential areas has probably contributed to their eventual cultivation, as they were not eliminated, favored its growth, robustness, and higher availability for consumption (McClung et al., 2014). According to the National Commission for the Knowledge and Use of Mexican Biodiversity, from the point of view of its agricultural management, amaranths do not represent competition for the main crops, since they are harvested or collected at the beginning of the agricultural cycle and these plants can be tolerated, encouraged, protected and cultivated in diverse agroecosystems such as milpas, natural vegetation, in associations with other plants or monoculture (Lozada and Neyra, 2021).

Amaranths are alternative crops, consumed throughout the Mexican territory (ASERCA, 2018) and are offered through traditional Mesoamerican markets as a wild resource and edible weed, an example of this is observed in 16 communities of Patzcuaro Lake and the Purépecha Plateau, places where amaranths are even produced in seasonal and irrigated agricultural plots (Farfán-Heredia et al., 2018). It is important to mention the cultural and gastronomic importance of amaranths in the central and southern regions of Mexico, where there are food guides, recipe books, dissemination material and amaranths are even part of dishes in prestigious traditional food restaurants; in contrast, in the north of the country, they are not so widely consumed and are even considered food for poor people. The gastronomy of amaranths in the central and southern regions of Mexico includes cymbals like: dip of pápalo and pipicha, pastes of verdolaga, quintonil candied, watercress salad, salad of amaranths, pumpkin guide soup, cannelloni stuffed with auazontle, chilpozonte of lenguas, egg with agrios, quintonil lasagna, fish with mole of chipilin, fish papalado, old woman's leg, amaranths with chorizo, cornfield roll, tortitas of chompantles, purslane with chicken, water of challa, water of chipilín and water of purslane (Delía et al., 2010).

In Mexico, more than 350 species of amaranths are consumed among its communities, these plants were selected over time by different peoples and regions. The frequently traded plants are papalo (*Porophylum ruderale* subsp.), purslane (*Portulaca oloracea*), romerito (*Suaeda nigra*), cenizo amaranth (*Chenopodium berlandieri* subsp. berlandieri), huauzontle (*C. berlandieri* subsp. nuttaliiae), alaches (*Anoda cristata*), epazote (*Dysphania ambrosioides*), chaya (*Cnidoscolus aconitifolius*), hoja santa (*Piper auritum*), chepiles (*Crotalaria* spp.) and quintonil (*Amaranthus* spp.) (Aranda and Neyra, 2021).

Amaranthus spp.

Plants known as amaranth comprise several families as the Amaranthaceae. This family is made up of 70 genera including Amaranthus which comprises 60 species that grow in Mexico. This family is known in the communities as huautli (Nahuatl language) and has been cultivated since pre-Hispanic times for its importance as food and a symbolic element of Mesoamerican culture. In addition, it was part of the ceremonies related to fertility and the agricultural cycle.

The above has been proven by archaeological studies carried out in various regions of the Mexican territory (Tehuacán, Puebla and Tlapacoyan) (SNICS, 2017; SEMARNAT, 2018; McClung and Martínez-Yrizar, 2017). In tropical America, where it is native, *Amaranthus* (meaning immortal in Greek) was a staple crop in Aztec, Mayan, and Inca civilizations. Currently, some of its species are cultivated and consumed in India, Nepal, China, Indonesia, Malaysia, Philippines, southern and eastern Africa, Central America and Mexico (Kavita and Puneet, 2017). The *Amaranthus* genus is being rediscovered as a food crop in different parts of the world, for its high nutritional value, both seeds and leaves. However, some of these species are considered edible, while others are still considered toxic, but, they have not been studied in-depth, for their food characteristics. In Cotonou, Republic of Benin, a study was conducted to better understand the evolutionary trend of these species and to develop an appropriate selection of *Amaranthus* lines; of the 30 descriptors found, the most used were related to characteristics such as inflorescence, floral parts, and seed traits, describing close relationships between plants, regardless of their place of origin (Sogbohossou and Achigan-Dako, 2014). Therefore, many of the plant species that are considered invasive, may serve as edible plants for human or animal consumption.

Table 1. Examples of species of *Amaranthus*

Amaranthus acanthochiton	Native Americans (Hopi) consumed the seeds and young leaves. The seeds were cooked as porridge and the leaves as vegetables with meat. The plant is considered a world heritage site in the G5 range, according to The Nature Conservancy (RHNM, 2021).
Amaranthus arenícola	The plant is native to the central and southwestern Great Plains. It is distributed from Texas to Nebraska or South Dakota and occasionally in North America and Europe. However, it is not domesticated (RHNM, 2021).
Amaranthus chihuahuensis	It was identified in November 1885 in the Corralitas locality of Chihuahua Mexico, (RHNM, 2021). However, it is unknown if it is edible or an invasive plant.
Amaranthus hybridus	The plant is widely distributed in Mexico, the Antilles, Central and South America. Additionally, it has been introduced in tropical, subtropical and warm temperate regions. The plant grows abundantly in agricultural fields and altered habitats and appears to hybridize with *A. retroflexus* (RHNM, 2021; CONABIO, 2009). (RHNM, 2021; CONABIO, 2009).
***Amaranthus hypochondriacus* L.**	Hybrids of this plant are widely cultivated as ornamentals, pseudo-cereals, and forages in many warm temperate and tropical regions worldwide.The initial cultivated form probably arose in southwestern North America, within the original range of *A. powelliinativo* (RHNM, 2021).
Amaranthus palmeri	The plant is native to southwestern North America and is distributed from southern California to Texas and northern Mexico. Currently, it is an invasive plant due to its spread to eastern North America. It blooms from summer-autumn, occasionally spring-winter (RHNM, 2021).
Amaranthus powellii	Introduced or naturalized in South America, Eurasia, Australia; native to the southwestern United States and adjacent regions of Mexico; now widely naturalized almost everywhere in temperate regions of North America. The distribution of *A. powellii* is probably underestimated (RHNM, 2021).
Amaranthus torreyi	The plant blooms during the summer-fall grows on sandy, rocky plains with gravel, slopes, canyons, washes and other natural habitats. The nomenclature and taxonomic relationships should be reviewed through experimental and field studies (RHNM, 2021).

Amaranthus retroflexus	The plant is native to central and eastern North America but was naturalized throughout most of the world. It is an invasive plant as it has effectively colonized a wide range of habitats on all continents. It exhibits wide variability and is easily recognized, Therefore easily identified (RHNM, 2021).
Amaranthus tricolor	It is a fast-growing annual plant, widely cultivated as a garden plant for its showy, often variegated, distal leaves of striking colors: red, scarlet, maroon, purple, yellow, cream and green. It is widely cultivated as an edible vegetable (CABI, 2016).
Amaranthus caudatus (Kiwicha)	It is another grain with similar characteristics to quinoa. However, its production is lower due to the lack of detailed knowledge of its nutritional properties and chemical composition. Recent data show that it contains fiber (11.59%), protein (2.48%) and carbohydrates (65.94%) (Velásquez-Barreto et al., 2021).
Amaranthus mantegazzianus	The plant contains protein obtained by different processes. It also contains phenolic compounds and antioxidant activity, so it could be used to make food (Castel et al., 2014).
Amaranthus muricatus	This wild plant grows well in poor soils unsuitable for grains. Nutritional potential of its leaves has been reported. Therefore, domestication of this plant is suggested together with the evaluation of its chemical and nutritional properties (Escudero et al., 1999).

Amaranthus growing conditions

There are different factors by which *Amaranthus* spp. are distributed worldwide but one of the most important is its resistance to biotic (diseases and pests) and abiotic (heat and drought) stresses (Bhargava and Srivastava, 2020). They also exhibit high tolerance to poor and saline soils, semi-arid climates, and unfavorable environmental conditions, so they have good biomass production and are dicotyledonous plants that produce grain-like seeds (SNICS, 2017; Idris et al., 2020). Their persistence in different climatic conditions is due to high phenotypic plasticity and genetic variability, with rapid growth rate, high fecundity, limited seed dormancy, and evolution of herbicide resistance (Sarangi et al., 2021). Physiologically these plants exhibit indeterminate flowering habits, abundant and lateral root system, solute accumulation, efficient water use and expression of stress-related genes and transcription factors (Bhargava and Srivastava, 2020). In India, 17 of these species have been identified with food-grade, for its part, the National Botanical Research Institute of India refers germplasm referable to 20 species (Kavita and Puneet, 2017); while in Mexico, research has been conducted aimed at the knowledge of the *Amaranthus* genus, its conservation and sustainable use, reporting 29 species of which 12 stand out among which are *A. hipochodriacus, A. caudatus, Amaranthus cruentus, A. hybridus* and *A. retroflexus* (SNICS, 2017; Achingan-Dako, Sogbohossou and Maundu, 2014). However, all the other mentioned species of the *Amaranthus* genus are considered among the most common weeds in the world and are subjected to different treatments to be eradicated (Tofiño et al., 2020).

For approximately ten years, an overview of plant health has been managed. These activities include a prioritization plan for pest risk analysis on invasive exotic plants, new standards, and regulations. Further development of cooperation between the environmental and plant health sectors. Actions that are important, because response of one plant to a stress situation is different to that of another plant, for example in competition for space and resources, alterations of the habitat where they grow, alterations due to parasitism or allelopathy, and even hybridization with cultivated plants or wild plants (Starfinger and Schrader, 2021). Therefore, researching, analyzing, and understanding the ability of *Amaranthus* species to adapt, develop and propagate in almost any type of soil and climate should generate the common interest of producers around the world. Investigating the physiological nature of the plant and its relationship with the environment in which it reproduces, could provide base knowledge for the care and adaptation of edible plants of interest for food security.

Some research on this point demonstrates the fact that coexistence between species can be favorable or negative for the plants of interest. In a study conducted in Pakistan was found changes in soil diversity after invasions of *Amaranthus viridis* L. (considered invasive and noxious in the United States and Canada), demonstrating the positive effect on the soil nutrient pool by increasing carbon, nitrogen, and phosphorus concentrations and microbial activities; the inoculation of arbuscular mycorrhizal (AM) fungi (Glomus intraradices) for the establishment of *A. viridis* was highly beneficial for growth and nodulation of *Acacia* species; while it negatively affected by altering the development of rhizobia and AM communities (Sanon et al., 2009). Agricultural habitats integrate a series of geological-morphological factors and processes in which vegetation responds in different ways, so analyzing soil deposits allows indicating the existing relationship between vegetation and its geomorphology, presenting a dynamic behavior that involves abiotic factors such as typographic, hydrological, lithological differences, etc. resulting in vegetation heterogeneity. Additionally, *A. standleyanus* growth has been detected as part of the native vegetation of geomorphological environments of the Central Arid Mountain of Argentina due to the active and dynamic processes in the fluvial and interfluvial areas of these zones (Flores et al., 2015). However, it does not explain the ability of *Amaranthus* spp. to reproduce and develop in extreme environments. One proposal is the leaf-water relationship which is characterized by a high capacity for osmotic adjustment, which can sustain the maintenance of turgor and, consequently, production during drought stress (Liu and Stutzel, 2002). As it can be observed, there are almost no studies that investigate synergy between *Amaranthus* species and the environment in which it develops, so it is necessary to focus simultaneous efforts on its understanding.

Nutrition and Food

The importance of wild edible plants collected and consumed as weeds (*Amaranthus* spp.) are signs of coping mechanisms for environmental and socio-cultural changes in a state of transition, where preferences are given according to socioeconomic, cultural, and ecological conditions, playing a role in mitigating food deficiency as food alternatives (Ong and Kim, 2017; Ntidi et al., 2019). In addition, the lifetime is short, so consumed in an early vegetative state since the plant has young leaves, soft stems and immature inflorescences. Therefore, the villagers consider them as vegetables. The main form of consumption of *Amaranthus* species is sewn as an accompaniment for dishes (with black beans, eggs, rice, quesadillas, sauces, soups, steamed, tamales, thickened, and covered with eggs) (Manzanero et al., 2020).

In the different species of *Amaranthus*, a great phenotypic variation is appreciated, and these plants can also have different nutritional attributes and bioactive compounds (Akinyi et al., 2021), ensuring nutrition-health needs and cultural identity (Mosyakin and Kenneth, 2020). In some places, they are recognized for having importance in traditional medicine (Kongdang et al., 2021; Schleiden et al., 2019). It is relatively recent when they have begun to give scientific relevance to the *Amaranthus* genus, studying some nutritional parameters including fatty acid profile, total protein content, amino acids, carotenoids, steroids, polysaccharides, azulenes, depsides, coumarins, and saponin; the bases of these data are based on *A. hypochondriacus* L. and some wild relatives such as *A. spinosus* L., *A. dubius* Mart. ex Thell., *A. hybridus* L., *A. tricolor* L., *A. cruentus* L., *A. caudatus* L., *A. retroflexus* L., etc.; (Shukla et al., 2018; Schleiden et al., 2019). On the other hand, the leaves contain oxalate, vitamin C, carotenoids and present morphological features are used as indicators of nutritional attributes (Akinyi et al., 2021).

They also contain phenolic acids and flavonoids (Kongdang et al., 2021). Therefore, they represent an alternative source of food to improve health. Additionally, *Amaranthus* plants contain tryptophan, sulfur amino acids, fibers, lipids rich in linoleic acids, carbohydrates, and minerals such as calcium, iron, magnesium, manganese, potassium, and sodium (Figure 1) (Regina et al., 2021).

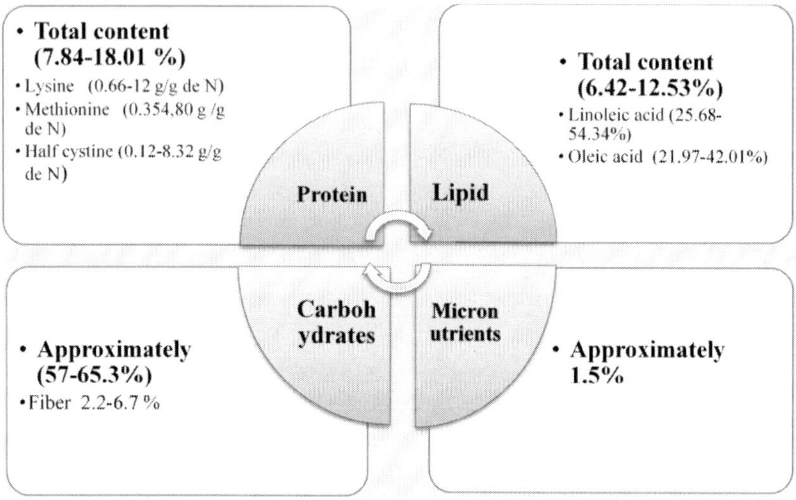

Figure 1. Nutritional composition of *Amaranthus* spp.
(Barros et al., 2021; Shukla et al., 2018).

Expectations

Currently, because climate crisis and biodiversity loss is necessary to understand complex indigenous agricultural systems as genetic reservoirs and identify synergies among productive landscapes, their conservation, and food security (Heindorf et al., 2021).

Alternative Nutrient Sources

Efforts around the world to improve dietary diversity must address food security challenges, support income generation and market access for diverse foods (Nabuuma et al., 2021). In the 2030 agenda, some goals embrace alternative foods as an important source of nutrients (protein) to improve dietary quality and implement non-conventional foods to combat food insecurity (Barros et al., 2021). Among the alternative foods, *Amaranthus* spp.

stand out not only for their nutrient quality, but also for being cultivated under adverse soil conditions (Idris et al., 2020). *Amaranthus* is a promising crop for semi-arid regions due to its high ability to adapt to drought stress, its biomass production, and water use efficiency. These plants can maintain a regulatory balance under water scarcity, significantly decreasing the total dry mass of the plant, differing the proportion between root, stem, and leaves. Interestingly, a negative relationship between specific leaf area and water use efficiency was observed in four genotypes under irrigated conditions (Liu and Stützel, 2004).

Amaranthus leaves are considered as vegetables and should be promoted as a direct source of nutrients and health compounds, a recommendation generated after the analysis of 15 *Amaranthus* species, where a serving of any of the leaves (1 cup; 30 g fresh weight basis) contributes as established by the U.S. Food and Drug Administration, 13 to 34% of the daily value (DV) of Mg and up to 68% of the DV of vitamin C. In addition, *Amaranthus acanthochiton* is considered a good source of Ca, Mn and Mo, and an excellent source of Mg and C vitamin (Jiménez-Aguilar and Grusak, 2017). In South Africa, *Amaranthus cruentus* is considered an edible leafy vegetable (Bello and Walker, 2017). The high phenotypic plasticity of *Amaranthus* species has allowed its consumption for many centuries in Africa, Asia, Central, and South America, making it a suitable crop for nutritional security in a changing world (Jamalluddin et al., 2021). On the other hand, the shortage of food gives rise to the search for alternatives to guarantee nutritional needs. Therefore, amaranths represent a food option with great nutritional value since they contain chlorophyll, β-cyanines, β-xanthines, betalains, carotene, ascorbic acid and antioxidants (Sarker et al., 2018).

Amaranthus species are considered potential crops with promising economic value, moreover, they could improve the biological value of processed foods, being one of the most promising vegetables to achieve a healthy life (Olaide et al., 2018). Grains of the *Amaranthus* genus have interesting amounts of lipids, finding a yield nine times higher in small particles (<0.23mm) concerning larger ones (>0.5mm). Squalene (4.3 - 6.0%), palmitic acids (16-26%), oleic (20-37%) and linoleic (37-47%) have been detected in some *Amaranthus* species (Kraujalis et al., 2013). The cold-pressed oil of *A. tricolor* is an example (Mondor, Melgar-Lalanne and Hernández-Álvarez, 2020).

Amaranthus has also been used to develop a film that contains leaf extract as an active ingredient. This film changes color and pH in the presence of microorganisms, which is why its use was proposed to preserve chicken and fish during refrigerated storage (Kanatt, 2020). On the other hand, some species of amaranth, such as *A. hypochondriacus* L. are considered pseudocereal. Therefore, its combination with *Bacillus pumilus* and *Bacillus subtilis* improves its nutritional value since it produces a natural, persistent effect and has long-lasting potential. Additionally, the bacillus promotes the increase of macronutrients (dry matter, proteins, fats and carbohydrates) and essential amino acids such as methionine, lysine and tryptophan. This confer them an advantage over major crops such as rice, wheat, and maize which are lacking in important nutrients such as essential amino acids, vitamins and minerals (Pandey et al., 2018). In Zagreb Croatia, nitrogen fertilization was implemented for two years in *A. hypochondriacus* L. and *A. cruentus* to generate more efficient production; as a result, seeds increased in percentage, quantity, and amount of protein (Pospisil et al., 2006). On the other hand, it has been mentioned that *Pseudomonas* spp. is another microorganism that can promote growth of *Amaranthus hybridus* and *Amaranthus viridis* when used in soil humus (Ikhajiagbe et al., 2021). Likewise, in Nepal, they are joining efforts for amaranth improvement programs, recommending stable and high-yielding genotypes to farmers. This, after identifying 13 *Amaranthus* genotypes subjected to stability analysis to determine their nature and extent of genetic stability (Kandel et al., 2021). From this type of plant, cookies have been prepared from raw amaranth grain flour and sprouts, the latter exhibited a higher number of antioxidants and total dietary fiber, being of acceptable quality and better nutrition (Chauhan et al., 2015).

Additionally, interest in the consumption of alternative cereals (pseudo-cereals) has recently increased. Therefore, *Amaranthus caudatus* has been studied to know the morphological changes that its grains undergo when are soaked at high temperatures. What gives rise to gelatinization of the starch at 60 ° C. In contrast, higher temperatures significantly altered texture and appearance of the grains. Amylose content correlates broadly with the functional properties of stickiness and texture (Lux et al., 2021; Kong et al., 2009). These data are very relevant for the food industry since the gelatinization temperature is low and therefore preserves its nutritional properties, with almost no loss due to the influence of the thermal process. The common name of *A. caudatus* is Kiwicha and it is considered as a multipurpose crop that supplies a rich source of protein (30% more protein with a complete set of amino acids), β-carotene, vitamins, minerals, and dietary fiber. Its starch is of promising use due to its high solubility and digestibility (it does not contain gluten), offering new possibilities for food processing, pharmacology, and cosmetics (Martinez-Lopez et al., 2020). It is worth mentioning that the nutritional properties of this plant change as it ages, having as an optimal cutting stage before flowering (Olaide et al., 2020).

Medicinal Potential

Plants are important natural sources of medicinal compounds. Since ancient times, plants have been used to treat various human diseases, and are associated with traditional medicine systems developed by recollection of plants in forests and other places. Medicinal properties of plants can be potentiated through a comprehensive understanding of molecular science and structural elucidation techniques (Rana et al., 2021). Several species of *Amaranthus* contain a combination of phytonutrients (particularly polyphenols) and dietary fiber, which may offer promising results in the treatment of certain conditions such as gastrointestinal disorders (Kongdang et al., 2021). Regular and adequate intake of phytonutrients offers protection against major chronic diseases (including cardiovascular, neurodegenerative, and cancer) while maintaining human health, some examples of this range are anthocyanins, carotenoids, coumarins, flavonoids, lignans, phenolic acids, polyphenols, sterols, and terpenes, which have been found in plant foods such as indigenous Andean crops (Repo-Carrasco-Valencia et al., 2010; Nahar et al., 2021). *Amaranthus viridis* has been shown to confer cardioprotective effects. After oral treatment of *A. viridis* for 45 days caused reduction of serum marker enzymes, cardiac troponin, and lipid peroxidation, raising the levels of antioxidant enzymes. The biological activity was higher when leaves were used in comparison when seeds of *A. hypochondriacus* were employed (Saravanan et al., 2013; Yawadio et al., 2008; López-Mejia et al., 2014). For its part, *Amaranthus retroflexus* is considered as a potential nutraceutical for humans because it has different phytochemical compounds, such as prenylated and non-prenylated phenylpropanoids (being in its roots where it is found the highest amount), umbelliferone, apigenin, 4′-geranyloxiferulic acid, 7-isopentenyloxycoumarin, auraptene, and umbelliprenin in all parts of this herb (Fiorito et al., 2017). The flavonoid rutin (quercetin-3-rutinoside) has health properties and benefits. It acts by regulating the immune system, can reduce allergies, as well as, offer resistance against bacterial and viral infections. This compound has been extracted from ground dried leaves of *Amaranthus paniculatus,* considering some factors such as extraction solvent, time, and temperature (Kraujalis et al., 2015). *A. caudatus* is highlighted as a potential source of biologically active compounds with antidiabetic, antihyperlipidemic, and antihypercholesterolemic effects and antioxidant and antimicrobial activities. Therefore, incorporation of *A. caudatus* seeds in the diet could be associated with health promotion and disease prevention. In general, these species are rich in polyphenols and dietary fibers. Several microbial metabolites are biologically active, so alteration of gut microbiota is largely related to the metabolic characteristic of plants. Based on the evidence available to date, several *Amaranthus* plants contain a combination of phytonutrients, particularly polyphenols and dietary fibers (Kongdang et al., 2021), which could help cope with a diet oversaturated with meats, cheeses and eggs. This increase of organic acids, while fruits and vegetables increase alkalis. Consequently some authors have suggested implementation of vegetarian diets as a strategy to reduce metabolic acidosis (Osuna-Padilla et al., 2019).

Ingredient in Agricultural Products

Studies carried out with *Amaranthus* leaves showed that these plants are rich in biologically active compounds with allelopathic effects capable of inhibiting germination, cell cycle, and initial development in seedlings (Schleiden et al., 2019). On the other hand, microorganisms were isolated from healthy roots with the ability to generate growth promotion in other plants and suppression of the foliar disease known as blight (Yashaswini, Nysanth and Anith, 2021). Another study demonstrates the botanical potential of these plants as a fungicide and extracts of *A. spinosus* were used to inhibit germination of spores of *Phakopsora pachyrhizi*, a causal agent of soybean rust disease. The methanol extract of roots and flowers inhibited approximately 55% of this fungus (Yusnawan, 2015).

Conclusion

Current information on *Amaranthus* species shows that this group of plants has biologically active components. Therefore, they represent a food alternative for humans; since they contain nutritional compounds to maintain human health. Additionally contains secondary metabolites with properties to prevent chronic diseases. This group of plants is resistant to adverse conditions. Therefore, it requires little care for its development. However, research is required to deepen the study of this group of plants to classify all species suitable for consumption and to know their adaptive growth capacity to establish the starting point to achieve technical-agricultural development in food security.

References

Achingan-Dako, E., Sogbohossou, O. & Maundu, P. (2014). Current Knowledge on *Amarnthus* spp.: research avenues for improved nutritional value and yield leafy amaranths in sub-Saharan Africa. *Euphytica*, *197*, 303-317.

Akinyi Nyonje, W., Schafleitner, R., Abukutsa-Onyango, M., Yang, R-Y., Makokha, A. & Owino, W. (2021). Precision phenotyping and association between morphological trair and nutritional content in Vegetable Amaranth (*Amaranthus* spp.). *Journal of Agriculture and Food Research 5*, *1000165*.

ASERCA. (update date December 14, 2018; consultation date October 17, 2021). *Service Agency of Agricultural Marketing and Market Development.* Single portal for procedures, information and citizen participation: https://www.gob.mx/aserca/articulos/que-son-los-quelites.

Barros Mariutti, L. R., Sarmento Rebelo, K., Bisconsin-Junior, A., Santos de Morais, J., Magnani, M., Rodrigues Maldonade, I., Rodrigo Madeira, N., Tiengo, A., Maróstica Jr, M. R. & Betim Cazarin (2021). The use of alternative food sources to improve health and guarantee access and food intake. *Food Research International*, Volume *149*, 110709.

Bello, Z. A. & Walker, S. (2017). Evaluating AquaCrop model from simulating production of amaranthus (*Amaranthus cruentus*) a leafy vegetable, under irrigation and rainfed conditions. *Agrucultural and Forest Meteorology, Volume*, *247*, 300-310.

Bhargava, A. & Srivastava, S. (2020). Response of Amaranthus sp. to Salinity Stress: Areview. In H. A., C.-A. R. and R. R., *Emerging Research in Alternative Crops. Environment and Policy*, Vol *58* (pp. 245-263). Springer, Cham.

CABI. (update date July 07, 2016; consultation date October 23, 2021). *Compendios de Especies Imbasoras*. Retrieved from: Department of Natural and Environmental Resources of Puerto Rico https://www.cabi.org isc/datasheet/112199#toPictures. [*Invasive Species Compendiums*]

Casas, A., Vázquez, M., Viveros, J. & Caballero, J. (1996). Plant management among the Nahua and Mixtec in the Balsas River Basin, Mexico: an ethnobotanical approach to the study of plant domestication. *Hum Acol*, *24*, 455-478.

Castel, V., Andrich, O., Netto, F. M., Santiago, L. G. & Carrera, C.R. (2014). Total phenolic content and antioxidant activyty of different streams resulting from pilot-plant processes to obtain *Amaranthus mantegazzianus* protein concentrates. *Journal of Food Engineering*, Volume *122*, 62-67.

Chauhan, A., Saxena, D. C. & Singh, S. (2015). Total dietary fibre and antioxidant activity of gluten free cookies made from raw and germinated amaranth

(*Amaranthus* spp.) flour. *LWT - Food Science and Technology*, Volume *63*, *Issue 2*, 939-945.

CONABIO. (update date October 16, 2009; consultation date September 06, 2021). *Amaranthaceae*, *Amaranthus hibridus L.* Retrieved from Weeds of Mexico: http://www.conabio.gob.mx/malezasdemexico/amaranthaceae/amaranthus-hybridus/fichas/ficha.htm.

CONABIO. (update date October 08, 2020; consultation date September 24, 2021). *National Commission for the Knowledge and Use of Biodiversity.* Retrieved from Mexican biodiversity: https://www.biodiversidad.gob.mx/biodiversidad/que_es.html.

Delía, C. L., Bye Boettler, R. & Mera Ovando, L. M. (2010). *Recipe of Quelítes of . Central and Southern Mexico.*México, DF: DR. National Autonomous University of Mexico.

Elouafi I., Ahmed Shahid, M., Begmuratov, A. & Hirich, A. (2020). The Contribution of Alternative Crops to Food Security in Marginal Environments. In Hirich A., Choukr-Allah, R. and Rabag R. (eds) *Emerging Research in Alternative Crops* (pp. 1-23). *Environment and Policy*, vol 58: Springer, Cham.

Escudero, N. L., Albarracín, G., Fernández, S., Arellanos, L. M. & Mucciarelli, S. (1999). Nutrient and antinutrirnt composition of *Amaranthus muricatus. Plant Foods Hum Nutr*, *54*, 327-336.

Farfán-Heredia, B., Casas, A., Moreno-Calles, A., García-Frapolli, E. & Castilleja, A. (2018). Ethneocology of the interchange of wild and mushrooms in Phurépecha markets of Mexico: economic motives of biotic resources management. *Journal of Ethnoviology and ethnomedecine*, *14*, 5.

Fiorito, S., Epifano, F., Palmisano, R., Genovese, S. & Taddeo, V. A. (2017). A re-investigation of the phytochemical composition of the edible herb Amaranthus retroflexus L. *Journal of Pharmaceutica and Biomedical Analysis.*, Volume *147*, 183-187.

Flores, D. G., Suvires, G. & Dalmasso, A. (2015). Distribution of native vegetation in Quaternary geomorphological environments of the Central Arid Mountain of Argentina. *Mexican Journal of Biodiversity*, Volume *86*, Issue 1, 72-79.

Guzmán-Lucio, M. A., Wendt, T., Simpson, B., Alvarado-Vázquez, M. A., Foroughbakhch-Pournavab, R., González-Álvarez, M. & Rocha-Estrada, A. (2013). Floristic list of winter-flowering annual species in northeastern Nuevo Leon Mexico. *Mexican Journal of Biodiversity*, Volume *84*, Issue 3, 884-893.

Heindorf, C., Reyes-Agüero, J., Fortanelli-Martínez, J. & Hooft, A. V. (2021). More than Maize, Bananas, and Coffee: The Inter-and Intraspecific Diversity of Edible Plants of Huastec Mayan Landscape Mosaics in Mexico. *Economic Botany*, https://doi.org/10.1007/s12231-021-09520-9.

Idris, L. M., Nulit, R., Qamaruz Zaman, F. & Arifin, F. K. (2020). Hydrotime analysis of amaranthus spp. seed germination under salinity condition. *Journal of Applied Research on Medicinal and Aromatic Plants, 17, 100249.*

Ikhajiagbe, B., Chidozie Ogwu, M., Olowaseun Fawehinmi, F. & Johnson Adekunle, I. (2021). Comparative growth responses of *Amaranthus* [L.] species in Humus and Ferruginous Ultisols using plant growth promoting Rhizobacteria (*Pseudomonas* species). *South African Journal of Botany*, Volume *137*, 10-18.

Jamalluddin, N., Symonds, R. C., Mayes, S., Kuan Ho, W. & Massawe, F. (2021). Champer 6 - Diversifying crops for food and nutrition security: Acase of vegetable amaranth, an ancient climate-smart crop. In Galanakis, C. M. *Food Security and Nutrition* (pp. 125-146). Academic Press.

Jiménez-Aguilar, D. M. & Grusak, M. A. (2017). Minerals, vitamin C, phenolics, flavonoids and antioxidan activity of amaranthus leafy vegetable. *Jornal of Food Composition And Analysis*, Volume *58*, 33-39.

Kanatt, S. R. (2020). Development of active/intelligent food packaging film containing *Amaranthus* leaf extract for shelf life extension of chicken/ fish during chilled storage. *Food Packaging and Shelf Life*, Volume *24*, 100506.

Kandel, M., Raj Rijal, T. & Kandel, P. (2021). Evualuation and identification of stable and yielding genotypes for varietal development in amaranthus (*Amaranthus Hipochondriacus* L.) under hilly of Nepal. *Journal of agriculture and food Research*, Volume *5*, 100158.

Kavita, P. & Puneet, G. (2017). Rediscovering the therapeutic potencial of *Amaranthus* species: A review. *Egyptian Journal of Basic and Appled Sciences*, Volume *4*, Issue 3, 196-205.

Kong, X., Bao, J. & Corke, H. (2009). Physical properties of Amaranthus starch. *Food Chemistry*, Volume *113*, Issue 2, 371-376.

Kongdang, P., Dukaew, N., Pruksakorn, D. & Koonrungsesomboo, N. (2021). Biochimistry of *Amaranthus* polyphenols and their potential benefits on gut ecosystem: A comprehensive review of the literature. *Journald of Ethnopharmacology*, Volume *281*, 114547.

Kraujalis, P., Rimantas Venskutonis, P., Ibáñez, E. & Herrero, M. (2015). Optimization of rutin isolation from *Amaranthus peniculatus* leaves by high pressure extraction techniques. *The Journal of Supercritical Fluids, Volume 104*, 234-242.

Kraujalis, P., Rimantas Venskutonis, P., Pukalskas, A. & Kazernaviciuté, R. (2013). Accelerated solvent extraction of lipids from *Amaranthus* spp. seeds and characterization of their composition. *LWT Food Science and Tecnology*, 528-534.

Liu, F. & Stützel, H. (2002). Leaf water relations of vegetable amaranth (*Amaranthus* spp.) in responseto soil drying. *European Journal of Agronomy*, Volume *16*, Issue 2, 137-150.

Liu, F. & Stützel, H. (2004). Biomass partitioning, especific leaf area, and water use efficiency of vegetable amaranth (*Amaranthus* spp.) in response to drought stress. *Scientia Horticulturae*, Volume *102*, Issue 1, 15-27.

López-Mejia, O. A., López-Malo, A. & Paluo, E. (2014). Antioxidant capacity of extracts from amaranth (*Amaranthus hipochondriacus* L.) seeds or leaves. *Industrial Crops and Products*, Volume *53*, 55-59.

Lozada Aranda, M. & Neyra, L. (update date May 19, 2021; consultation date November 11, 2021 2021). *Mexican biodiversity*. Retrieved from Natural and cultural diversity: https://www.biodiversidad.gob.mx/diversidad/alimentos/quelites.

Lux, T., Wernecke, C., Bosse, R., Reimold, F. & Flöter, E. (2021). Textrural and morphological changes of heat soaked raw *Amaranthus caudatus*. *Journal of Cereal Science*, Volume *98*, 103168.

Manzanero Medina, G. I., Vásquez-Dávila, M., Lustre-Sánchez, H. & Pérez-Herrera, A. (2020). Ethnobotany of food plants (quelites) sold in two traditional markets of Oaxaca, Mexico. *South African Journal of Botany*, 215-223.

Mapes, C. & Barsuto, F. (2016). Biodiversity and Edible Plants of Mexico. In Lira, R., Casas, A. and Blancas, *J. Ethnobotany of Mexico* (pp. 83-131). New York, NY: Springer.

Martinez-Lopez, A., Millán-Linares, M. C., Rodríguez-Martín, N. M., Millán, F. & Montserrat de la Paz, F. (2020). Nutraceutical Value of Kiwicha (*Amaranthus caudatus* L.). *Journal of Functional Foods*, Volume *65*, 103735.

McClung de Tapia, E., Martínez Yrízar, D., Ibarra Morales, E. & Adriano Morán, C. C. (2014). The pre-Hispanic origins of a food tradition in the Mexico Basin. *Annals of Anthropology*, Volume *48*, 97-121.

McClung de Tapia, E. & Martínez-Yrizar, D. (2017). The potential of paleoethnobotanical evidence for the study of Totihuacan foodways. *Archeological and Anthropological Sciences*, *9*, 39-50.

Mondor, M., Melgar-Lalanne, G. & Hernández-Álvarez, A. J. (2020). Chapter 11 - Cold pressed amaranth (*Amaranthus tricolor*) oil. In Ramadan, M. F. *Cold Pressed Oild* (pp. 13-127). Academic Press.

Mosyakin, S. L. & Kenneth R, R. (Consultation date October 17, 2021). *Fora of North america.* Retrieved from FNA vol. 4 | Amaranthaceae: http://www.efloras.org/florataxon.aspx?flora_id=1&taxon_id=101257.

Nabuuma, D., Ekesa, B., Faber, M. & Mbhemyane, X. (2021). Community perspectives on food security and dietary diversity among rural smallholder farmers: A qualitative study in central Uganda. *Journal of Agriculture and Food Research*, 100183.

Nahar, L., Xiao, J. & Sarker, S. D. (2021). Introduction of Phytonutrients. In Xiao, J., Sarker, S. D. and Asakawa, Y. *Handbook of Dietary Phytochemicals* (pp. 1-17). Singapur: Springer.

Ntidi, K., Daneel, M. & Fourie, H. (2019). The host status of *Amaranthus* spp. genotypes to *Meloidogyne* spp. and verification of resistance in genotype Thohoyandou. *Crop Protection*, Volume *115*, 64-74.

Olaide Jimoh, M., Jide Afolayan, A. & Bayo Lewu, F. (2018). Suitability of *Amaranthus* species for alleviating human dietary dificiencies. *South African Journal of Botany*, Volume *115*, 65-73.

Olaide Jimoh, M., Jide Afolayan, A. & Bayo Lewu, F. (2020). Nutrients and antinutriendt con stituents of *Amaranrhus caudathus* L. Cultivated on different soils. *Saudi Journal of Biological Sciences*, Volume *27*, Issue 12, 3570-3580.

Ong, H. G. & Kim, Y. (2017). The role of wild edible plants in household food security among transitioning hunter-gatherers: evidence from the Philippines. *Food Security*, *9*, 11-24.

Osuna-Padilla, I. A., Leal-Escobar, G., Garza-García, C. A. & Rodríguez-Castellanos, F. (2019). Dietary acid load; mechanisms and evidence of its impact on health. *Nephrology*, Volume *39*, Issue 4, 343-354.

Pandey, C., Bajpai, V. k., Kumar Negi, Y., Rather, I. A. & Maheshwari, D. K. (2018). Effec of plant growth promoting Bacillus spp. on nutritional propierties of *Amaranthus hypochondriacus* grains. *Saudi Journal of Biological Sciences*, 1066-1071.

Picó, B. & Nuez, F. (2000). Minor crops of Mesoamerica in early sources (I). *Genetic Resources and Crop evolution*, *47*, 527-540.

Pospisil, A., Pospisil, M., Varaga, B. & Svecnjak, Z. (2006). Grain yieldprotein concentration of two amaranth species (*Amaranthus* spp.) as influenced by the nitrogen fertilization. *European Journal of Agronomy*, Volume *25*, Issue 3, 250-253.

Ramos de Robles, S. L., Garibay-Chávez, G. & Curiel-Ballesteros, A. (2019). Identification, collection and consumption of weeds and wild vegetables in Mexican communities: institutionalized local ancestral indigenous knowledge as ecological literacy, place and identity. *Cultural Studies of Science Education*, *14*, 1011-1030.

Rana, A., Anand, J., Tyagi, M. & Rai, N. (2021). Forest-Based Medicinal Plants for Cancer Cure. In: Husen, A., Bachheti, K. and Bachheti, A. *Non-Timber Forest Products* (pp. 255-280). Springer, Cham.

Repo-Carrasco-Valencia, R., Hellström, J. K., Pihlava, J. M. & Mattila, P. H. (2010). Flavonoids and other phenolic compounds in Andean indigenous grains: Quinoa (*Chenopodium quinoa*), kañiwa (*Chenopodium pallidicaule*) and Kiwicha (*Amaranthus caudatus*). *Food Chemistry*, Volume *120*, Issue 1, 128-133.

RHNM. (consultation date Octobert 12, 2021). *Herbarium Network of Northeastern Mexico.* Retrieved from https://herbanwmex.net/portal/taxa/index.php?taxon=Amaranthus+retroflexus&formsubmit=Search+Terms

Sanon, A., Béguiristain, T., Cébron, A., Berthelin, J., Ndoye, I., Leyval, C., Sylla, S. & Duponnois, R. (2009). Changes in soil diversity and global activities following invasions of the exotic invasive plant, *Amaranthus viridis* L., decrease the growth of native sahelian *Acacia* specie. *FEMS Microbiology Ecology*, Volume *70*, Issue 1, 118-131.

Sarangi, D., Jhala, A. J., Govindasamy, P. & Brusa, A. (2021). Charper 2 - *Amaranthus* spp. In Chauhan, B. S. *Biology and Management of Problematic Crop Weed Species* (pp. 21-42). Academic Press.

Saravanan, P., Ponmurugan, P., Sathiyavathi, M., Vadivukkarasi, S. & Sengottuvelu, S. (2013). Cardioprotective activity of *Amaranthus viridis* Linn: Effect on serum marker enzymes, cardiac troponin and antioxidant system in experimental myocardial infarcted rats. *International Journal of Cardiology*, Volume *165*, Issue 3, 494-498.

Sarker, U., Tofazzal Islam, M., Golam Rabbani, M. & Oba, S. (2018). Variability in total antioxidant capacity, antioxidant leaf pigments and foliage yield of amaranth. *Journal of Integrative Agriculture*, Volume *17*, Issue 5, 1145-1153.

Schleiden SousaCarvalho, M., Andrade-Vieira, L. F., dosSantos, F. E., Folgaroli Correa, F., Graças Cardoso, M. d. & Resende Vilela, L. (2019). Allelopathic potential and phytochemical screening of athanolic extracts from five species of *Amaranthus* spp. in the plant model *Lactuca sativa*. *Scientia y Horticulturae*, volume *245*, 90-98.

SEMARNAT. (update date October 31, 2018; consultation date November 13, 2021). *Ministry of Environment and Natural Resources.* Retrieved from Amaranth, deep-rooted and profusely rooted plant in Mexico: https://www.gob.mx/semarnat/articulos/amaranto-planta-de-profunda-y-profusa-raiz-en-mexico?idiom=es .

Shukla, A., Srivastava, N., Suneja, P., Yadav, S. K., Hussain, Z., Rana, J. C. & Yadav, S. (2018). Untapped Amaranth (*Amaranthus* spp.) genetic diversity with potential for nutritional enhancement. *Genetic Resources and Crop Evolution*, *65*, 243-253.

SNICS. (update date August 01,2017; consultation September 28, 2021). *National Seed Inspection and Certification Service.* Retrieved from Basic and Industrial Macro Network. Amaranth Network Overview: https://www.gob.mx/snics/acciones-y-programas/amaranto-amaranthus-spp.

Sogbohossou, O. E. & Achigan-Dako, E. G. (2014). Phentic differentiation and use-type delimitation in *Amaranthus* spp. from worldwide origins. *Scientia Horticulturae*, Volume *178*, 31-42.

Starfinger, U. & Schrader, G. (2021). Invasive alien plnts in health revised: another 10 years. *EPPO Bulletin*. epp 12787.

Tofiño Rivera, A. P., Carbono Murgas, R. E., Melo Ríos, A. E. & Merini, L. J. (2020). Effect ofglyphosate on microbiota, soil quality and biofortified bean crop in Codazzi, department of Cesar, Colombia Vol. *52*. *Argentine Journal of Microbiology*, 61-71.

Velásquez-Barreto, f. F., Artega Miñano, H., Alvarez-Ramirez, J. & Bello-Pérez, L. (2021). Structural, functional, and chemical properties of small starch granules: Andean quinoa and Kiwicha. *Food Hidrocolloids*, Volume *20*, 106883.

Vibras, H. (2016). Ethnobotany of Mexican Weeds. In Lira, R., Casas, A. &. Blancas, *J. Ethnobotany of Mexico. Ethnobiology* (pp. 287-317). New York, NY.: Springer.

Yashaswini, M. S., Nysanth, N. S. & Anith, K. N. (2021). Endospore-forming bacterial endophytes from *Amaranthus* spp. improve plant growth and sujppress leaf blight (Rhizoctonia solani Kühn) disease of *Amaranthus tricolor* L. *Rhizosphere*, *19*, 100387.

Yawadio Nsiba, R., Kikuzaki, H. & Konishi, Y. (2008). Antioxidant activity of various extracts and fractions of *Chenopodium* quinoa and *Amaranthus* spp. seeds. *Food Chemistry*, Volume *106*, Issue 2, 760-766.

Yusnawan, E. (2015). Inhibition of Spore Germination of Phakopsora Pachyrhizi Using Crude Extracts of *Amaranthus spinous*. *Procedia Food Science*, Volume *3*, 340-347.

Chapter 5

Refrigeration Systems and Their Application in the Food Industry

Ninja Begum[1,*] and Sanjib Kr. Paul[2,†]
[1]Department of Food Science and Technology,
University of Science and Technology, Meghalaya, Meghalaya, India
[2]Business Development and Research Management Division,
CSIR-Indian Institute of Chemical Technology, Hyderabad, India

Abstract

Refrigeration also known as artificial cooling is the process of exchanging heat from a source (an object) at high temperature to a sink at lower temperature obeying second law of thermodynamics. This chapter describes the fundamentals of refrigeration, its types, recent developments and multi-dimensional applications in food industries. The vapor compression cycle is discussed in details describing all its four components (evaporator, compressor, condenser & expansion valve) and how refrigerants flow between these components changing its state to produce refrigeration effect. This cycle is also explained with the help of pressure-enthalpy and temperature-entropy diagrams. This cycle gives an insight into how cooling is achieved at its maximum level while minimizing the amount of work done. Further, this chapter presents the different types of refrigerants and the way of designating those refrigerants. Also the operating concept of vapor absorption refrigeration circle and multistage refrigeration system is succinct emphasizing its

* Corresponding Author's E-mail: ninzasworld@gmail.com.
† Corresponding Author's E-mail: sankumarpaul@gmail.com.

In: Food Processing
Editors: Myriam Huijs
ISBN: 978-1-68507-570-5

distinctions cascade refrigeration system. This chapter also touches various household and industrial applications of these systems for the purpose of storage and post-harvest management of food materials with technical elaborations.

Keywords: compressor, condenser, evaporator, expansion valve, refrigerants, refrigeration, food storage

Introduction

In this chapter refrigeration, its types and its working are explained in details. Refrigeration is a preservation technique in which foods are kept below ambient temperature retarding the growth of microorganisms increasing their shelf life [1]. Since growth of both living organisms depend largely on the physical environment. It plays a great role in slowing down the deterioration of food products which is due to biological activities and chemical processes, retaining the quality of food. Mostly eatables survive longer at colder temperature compared to normal temperature. Hence, perishable foods can have extended *storage life* for several days under refrigerated condition [2]. Thus, study of refrigeration is an important topic in the area of Food Science and Technology. The refrigeration system is an integration of modules connected sequentially to produce a cooling effect. The cooling effect is achieved by transferring heat from a sink to the source where low-temperature is desired. Thus fundamental of refrigeration system is to provide cooling effect to the food stored so that it could retain its nutritional value for a longer duration of time when kept inside a closed chamber meeting its coveted degree, which is lower than our surrounding. Specifically refrigerators control the growth of microbes producing suitable environment for extending its shelf life [3]. A household refrigerator is one of the best examples of a refrigeration system. Refrigerator enables food products stay for longer duration maintaining temperature, humidity, pressure, and air motion of inside compared to external environment. These variables are controlled mechanically by means of two cycles vapor compression and vapor absorption. During these cycles the above mentioned variables undergo physical changes producing cooling effect. Refrigeration can be said one of the best techniques for preservation of fruits and vegetables along with various other food products. Working of a refrigerator can be compared to a heat pump in terms of flow of heat. Likewise, refrigerator withdraws heat from a cold reservoir and dumps it to edibles kept for cooling. And for doing that refrigerator needs a pump obeying the second law of thermodynamics. Since the transfer of heat from colder to warmer region is non spontaneous, hence external work i.e., use of heat pump is essential. This dissipation of heat by artificial means is called refrigeration which is obtained by mechanical means pumping refrigerant through the four basic components-evaporator, compressor, condenser and expansion valve connected in a cyclic fashion [4]. For producing refrigeration effect, a working fluid known as refrigerant flows through each component changing its state from liquid to vapor and vice versa by the process of evaporating and condensing itself. While evaporating, the refrigerant absorbs latent heat and during condensing refrigerant gives out latent heat. Till date a number of refrigerants have come into account such as ammonia, CFCs, HCFCs, HFCs etc. It was during 17th century William Cullen of University of Glasgow in 1748 introduced artificial cooling or mechanical refrigeration. Cullen's model was further enhanced by American scientists Oliver Evans in 1805 and John Gorrie in 1844 followed by more modifications over years. With each passing years and each passing invention, refrigeration gained more popularity finding its way in household refrigerators, air conditioning, industrial freezers, cryogenics and so on [5]. Refrigeration systems have impact in both domestic and industrial application. When it comes to industrial application ultra low temperature is also required which is achieved by multistaging the components such as evaporator or compressor [6]. Compared to single stage refrigeration which produces cooling up to -30°C multi stage can produce below -60°C. Thus multistage along with single stage has numerous applications in food industry. Different products need different storage temperature and depending on that the chamber is designed. For fresh agricultural produce such as apple, lemon, carrot etc. the storage temperature is kept higher than that of meat and fish products. In some cases the chamber is provided with gases which is modified or controlled from outside. Thus in context of industrial application this device possess multifaceted applications in domestic as well as various food and allied industries, including – medicines and chemicals, bakery, dairy, brewing plants, fish processing units, meat industries and many more.

Common Terminologies Used in Refrigeration System

Some of the most important terminologies used for refrigeration system are enlisted in Table1.

Table 1. Common terminologies used in refrigeration system

Terminology	**Description**
Enthalpy	Enthalpy is often regarded as a system's total heat content. It can be calculated as the sum of internal energy and the product of volume and pressure.
Entropy	Entropy is a measure of the energy dispersal in the system.
Latent Heat	It is the amount of heat required to change the state of a matter keeping the temperature constant.
Sensible Heat	It is the heat that causes a change in temperature in an object by no change in state.
Specific Heat	It is the amount of heat required to raising the temperature by one degree Celsius divided by total mass of the substance.
Thermal conductivity	*Thermal conductivity* can be defined as the rate at which *heat* is transferred by conduction through unit cross-section area of a material, when a temperature gradient exits perpendicular to the area. It is derived from the formula, Q=kAdT/dX
Thermal diffusivity	Thermal diffusivity is thermal conductivity divided by density and specific heat capacity at constant pressure. It is derived from the formula, $\alpha=k/\rho C_p$
Natural Convection	Natural flow of fluid where motion is not generated by any external source is called natural convection.
Forced Convection	Whenever fluids are forced to flow with the aid of an external force in order to increase the heat transfer it is known as forced convection.
Heat Exchanger	A *heat exchanger* involves transfer of *heat* between two or more fluids.
Sub cooling	When a liquid exists at a temperature below its normal boiling point that condition is known as sub cooling.
Super heating	*When* a liquid is heated to a temperature higher than its boiling point, without boiling is known as superheating.

Types of Refrigeration System

Refrigeration systems can be broadly classified into two categories. The two categories are - *Natural Refrigeration System* and *Mechanical Refrigeration System.* Before the invention of mechanical refrigeration in mid 19th century, preservation by artificial means was much awaited and till then natural means such as ice, snow, cold nights and cool caves, pits were dug into the ground and insulated with wood and straw. Ice that was used was transported from distant mountains and to store that ice proper insulation has been used. This system was then used for preserving food. Since pre-historic era cooling was the technique followed to preserve perishables for months and even years with minimal loss of nutritional value or change in flavor and appearance. The above techniques were just a savior till mechanical means came into effect.

Natural Refrigeration System

As discussed above, before the invention of mechanical refrigeration, eatables were preserved using natural processes, primarily by ice. What they did was they kept ice in an insulated chamber where the food was kept. Now to melt this ice to water, latent heat of 333.2 kJ/kg was required and this heat is extracted from the food that is being stored for preservation. From here comes the definition of Ton of Refrigeration (TR). One Ton of Refrigeration implies latent heat of one tonne of ice or the heat required to melt one tonne (2,000 lbs) of ice in pure state at 0°C (or 32°F) in 24 hours.

$$1\,TR = \frac{Weight\ of\ 1\ tonne\ of\ ice \times latent\ heat\ of\ melting\ ice}{24\ hrs}$$
$$= \frac{2000\ lbs \times 144\ Btu/lbs}{24\ hrs}$$
$$= 12000\ Btu/hr$$

This is the definition of 1 ton of refrigeration. The Capacity of a refrigeration unit is expressed in terms of Ton.

Mechanical Refrigeration System

Mechanical refrigeration gained its utility in the way that this system is operated mechanically for cooling stored products contained in it. Two of its most common types are: vapor compression refrigeration system (VCRS) and vapor absorption refrigeration system (VARS).

VCRS technique is extensively used in transferring heat to a higher temperature substance from a low temperature source by compressing the refrigerant flowing through its components. While, VARS involves the absorption process of the refrigerant by a transport medium called absorber producing refrigeration effect by thermal energy input. In this chapter, we will emphasize more on VCRS.

Vapor Compression Refrigeration System (VCRS)

VCRS procedure is most widely used among both the refrigeration systems. As its name, here vapor cycles are involved. It uses a circulating refrigerant that undergoes phase change until it administers its correct pressure, which involves the refrigerant turning into vapor from liquid and vice versa. In a vapor compression refrigeration system, cooling effect is attained as the flowing refrigerant evaporates at low temperatures [7]. This system operates through mechanical process. Hence it is called as mechanical refrigeration systems. The five main components of a VCRS as shown in Figure 1 are:

- Evaporator
- Compressor
- Condenser
- Expansion Valve (Throttling device)
- Refrigerant

Evaporator

Evaporator is commonly known as the 'heat exchanger' of the system. The evaporator is usually placed at the inside of the refrigerator. It is actually the component that accounts for refrigeration effect induced by the system [8]. The refrigerant which is at a low pressure and low temperature enters the evaporator. This flowing refrigerant when passes through the evaporator comes in contact with the zone of stored products present inside the refrigerator. This phenomenon i.e., flowing of refrigerant through the evaporator coils causes absorption of heat by it. After absorbing this latent heat the liquid refrigerant flowing through the evaporator coils transforms into vapor and retires from the evaporator and get into the compressor in vapor state [9]. Hence it is clear that evaporator involves heat transfer phenomenon and thus can be classified in accordance with the mode of heat transfer as- a. Natural convection also known as free convection evaporator, and b. Forced convection evaporator. Natural convection currents occur through natural process arises due to density difference caused by temperature difference. Whereas a fan or a pump is used in case of forced convection type evaporator to forcefully circulate the refrigerant to achieve rapid cooling effect. Evaporator is again categorized two general types i.e., 'direct or dry expansion' evaporator and 'flooded' evaporators. In 'flooded' evaporator, the refrigerant is not fully evaporated and heated, while in a direct expansion evaporator, the refrigerant is fully evaporated and heated at a high temperature. Table2 presents the evaporator types and its functions.

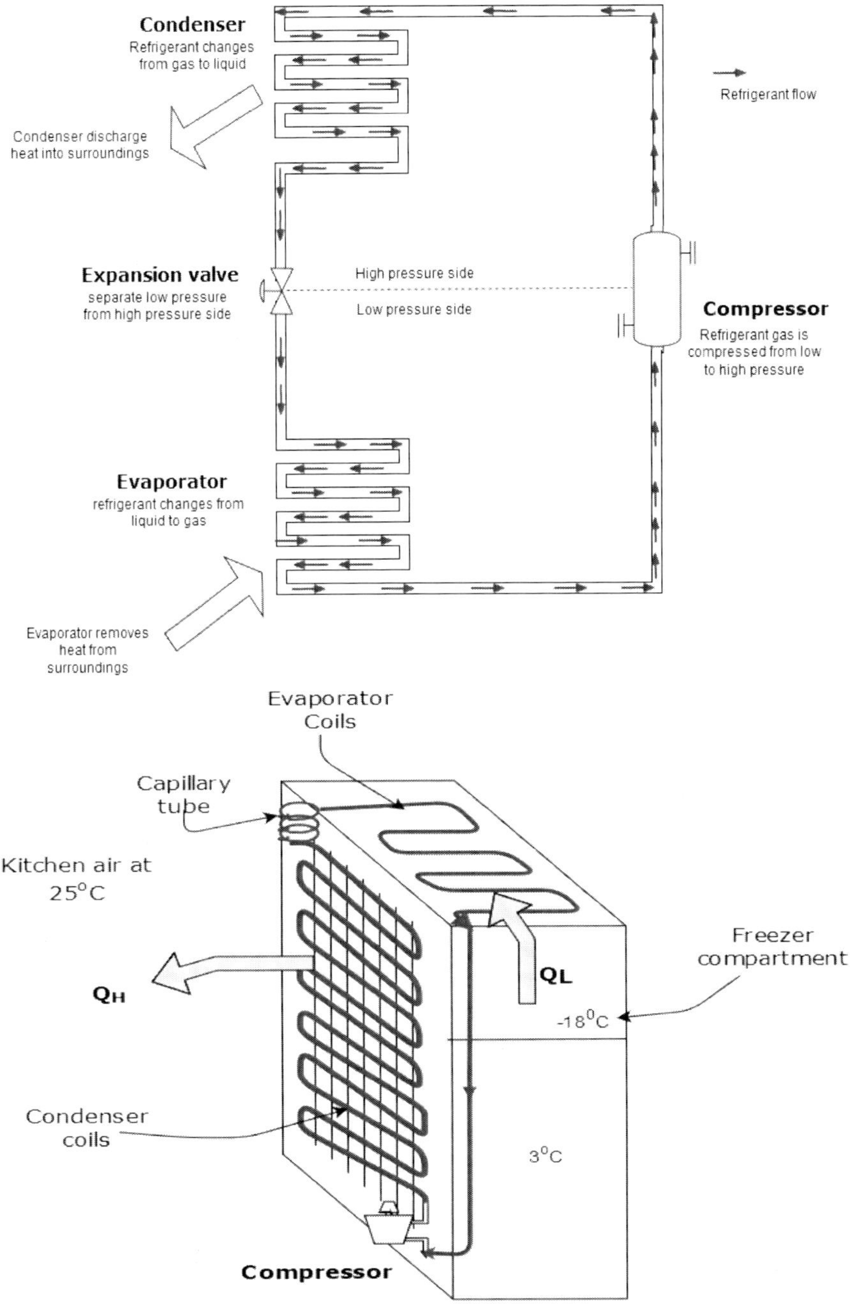

Figure 1. Schematic representation of Vapor Compression Refrigeration System.

Table 2. Types of evaporators with functional descriptions

Evaporator types	Description
Natural convection	It consists of short (5-10 cm) and vertical (1-2 m) tubes placed inside a steam chamber. Together this is called calandria. The refrigerant flows inside due to density difference. Here the velocity of fluid is low.
Forced convection	Here the fluid flows at a high velocity. A pump is used to achieve high circulation rates.
Dry expansion	The liquid refrigerant entering into the evaporator is converted to superheated vapor and is fully evaporated.
Flooded' evaporator	Amount of refrigerant supplied to this evaporator depends on the load on evaporator. Such evaporator is not always filled with the fluid. When load increases superheat vapor increases and hence more refrigerant flows as sensed by filler bulb.

Compressor

Compressor is the heart of the refrigeration system in the sense that it adds pressure to the refrigerant that flows through it. The vaporized refrigerant from the heat exchanger goes inside the compressor, where it is compressed adiabatically to very high pressure. This results in the rise of the temperature of the refrigerant [10]. Electrically operated compressors that are in use are of various types i.e., (a) Reciprocating compressor (b) Centrifugal compressor (c) Rotary compressor (d) Screw compressor (e) Scroll compressor. Reciprocating compressor compresses the vapor refrigerant by reciprocating motion of the piston. Centrifugal compressor produces centrifugal force and consists of an inlet casing, an impeller, a diffuser and a volute casing. Rotary compressors fall into positive displacement category. Uses of other two types are very limited. Compressor and its types are explained in Table3.

Condenser

The condenser is also known as the *freezer* of the refrigeration system. Like evaporator, the condenser is also a heat-exchange equipment. The condenser receives the refrigerant flowing out of the compressor. It is primarily used to cool down the refrigerant by absorbing the heat using cool air or water flowing through the condenser. In this process of cooling, the vaporized refrigerant condenses to liquid when the temperature is lowered but the pressure remains constant. Thereby this process is isobaric. Condensers are of three types, viz.

(a) Air-cooled Condenser (b) Water-cooled Condenser (c) Evaporative Condenser. Air cooled condensers are of two types (i) natural convection (ii) forced convection. Water cooled is of three types (i) tube-in-tube (ii) shell-and-coil (iii) shell-and-tube. All of the condensers are explained in

Table 4. Evaporative condensers are mixture of both air and water cooled types where condensing medium is both air and water.

Table 3. Types of compressor with functional descriptions

Compressor types	**Description**
Reciprocating	Reciprocating compressor uses piston driven by motors to compress the refrigerant.
Centrifugal	An impeller rotates inside the housing producing centrifugal force.
Rotary	These have screws, vanes, scrolls etc. which rotates and compresses the fluid.
Screw	In screw compressor, a pair of helical rotors rotates around a cylinder inside which fluid is trapped and compressed.
Scroll	It has two scrolls one stationary and the other moving. Upon rotation fluid is compressed.

Table 4. Types of condenser with functional descriptions

Condenser types	**Description**
Air- cooled	'Air Cooled' condenser is used where cooling load is low. In natural air cooled condensers, air flows in a natural method relying on the condition of condenser coil whereas, in 'forced air cooled' condenser a motor which operates fan is used for cooling.
Water-cooled	'Water Cooled' condenser is used where cooling load is high. In tube in tube condenser water moves through the tube and refrigerant moves through the circular space between the tubes. The direction of its motion is opposite to each other for maximum heat transfer. In shell and coil condenser water moves through the coil and refrigerant in shell exchanging heat between them. In shell and tube, hot vapor refrigerant enters the shell and gets condensed by water present inside the coils.
Evaporative	It accommodates both air cooled and water cooled condenser in the way that both air and water is used.

Expansion Valve

Finally, the refrigerant from the condenser which is at high pressure undergoes a sudden reduction in pressure in the expansion valve. The refrigerant further experiences decrease in temperature (sub-zero level) with the lowering of pressure inside the compressor. The refrigerant from this module then goes inside the evaporator and the cyclic sequence continues [11]. Expansion Valves are of various types, of which the most usual ones are: (a) Capillary tube, (b) Thermostatic expansion valve which is explained in Table5.

Table 5. Types of expansion valve with functional descriptions

Expansion valve types	**Description**
Capillary tube	In place of orifice, it consists of small diameter 0.5 to 2.28 mm and long length tubes wounded into coils where refrigerant is expanded by throttling.
Thermostatic	It expands as well as controls the refrigerant flowing into and out of the expansion valve depending on the load requirement.

Refrigerants

Refrigerant is defined as a substance or mixture of different substances, usually a gas at room temperature that can undergo phase transitions to produce a refrigeration effect. Some common refrigerants are ammonia, sulfur dioxide, etc. Chemical refrigerants are often assigned an R number which can be obtained systematically according to molecular structure. R number is the systematic method of representing the refrigerants. R number consists of the letter 'R' followed by two or three digits. Thus the representation RXYZ gives the chemical formula of the refrigerant used. One of the most important points to be taken care of about refrigerant is its compatibility with the environment [12]. Refrigerant must be non toxic, inflammable and non harmful. In such cases natural refrigerants can be an alternative [13].

Working principle of refrigerants: Refrigerants are one of the most important components that flow through the other components of the refrigeration system. While flowing through the components it absorbs as well as releases heat to produce refrigeration effect. Evaporator is where heat is absorbed and condenser is the component where heat is released. During this process of absorption and release of heat it changes is state from liquid to gas and vice versa [14].

Types of Refrigerants

A wide variety of refrigerants are used in vapor compression system. Refrigerant should have favorable thermodynamic as well as chemical properties for smooth functioning of the system. As refrigerant is a key component that flows through all the components so selection of refrigerant becomes challenging. Challenge not only lie in reducing emissions in the process, but also in making it an equitable choice to have a pragmatic and positive effect on functional efficiency, which reduces operating cost.

Based on working principle, refrigerants are classified as:

- *Primary:* Primary refrigerants are those which produce refrigeration effect by absorbing latent heat within the system. E.g., NH_3, SO_2 etc.
- *Secondary:* Secondary refrigerant exchanges heat within the system maintaining the temperature inside. It also reduces the harmful effect caused by primary refrigerants. Unlike primary refrigerant, it does not undergo phase change. E.g., brine

Based on chemical composition, refrigerants can be classified into the following categories:

- Fully saturated, halogenated compounds
- Inorganic refrigerant
- Mixtures
- Hydrocarbons (HCs)

Fully Saturated, Halogenated Compounds

It belongs to the family of refrigerants consisting of fluorinated and/or chlorinated hydrocarbons. These refrigerants are derived from the class of alkanes (C_nH_{2n+2}). This class of refrigerants is identified by a number assigned to them, which is given as R $X_CY_HZ_F$. Here, X_C, Y_H and Z_F denotes the count of carbon (C) atoms minus one, Hydrogen (H) atoms plus one, and fluorine (F) atoms respectively. The count of chlorine (Cl) atoms is obtained from the balance formula where the number of hydrogen and fluorine atoms is subtracted from the total number of atoms in the chemical formula. Sometimes, only two digits follow after R then the value of X_C is considered to be zero, e.g., R21. Here, X_C=0 which implies that there is 1 carbon atom. Y_H is 2 which indicate that the number of Hydrogen atoms is 1 and Z_F =1 indicates 1 fluorine atom. Number of chlorine atoms is given by {3-(1+1)}=1. Therefore, the chemical formula of R21 is *CHFCl.*

Inorganic Refrigerant

Refrigerant composed of inorganic compounds such as ammonia, sulfur dioxide, etc. which is designated using the molecular weight of the compound and is preceded by 7. For example, Ammonia (NH_3) is designated as 717 as its molecular weight is 17.

Mixtures

They are classified into two types: Azeotropic mixture (500 series) and Zeotropic mixture (400 series). In *azeotropic mixture,* substances having similar behavior but different chemical properties are mixed together. For example, R-502: 48.8% of R22 is mixed with 51.2% of R115. On the other hand, zeotropic mixture or blends are produced by combining substances of varying boiling points. Thus, in the refrigeration cycle, different substances within the mixture do not evaporate or condense at the same temperature. An example of zeotropic mixture is R404: 44%, 52% and 4% of R125, R143, and R134 respectively are mixed.

Hydrocarbons (HCs)

Hydrocarbon based refrigerants are organic chemical compounds which are entirely composed of hydrogen and carbon atoms such as ethane, methane, propane, etc.

Properties of Refrigerants

Properties of substances are one of the most important parameters that make a substance suitable as refrigerant. Following are the important properties of Refrigerants:

- Non-flammable: Refrigerants should be selected in such a way that it does not show flame propagation under certain tests.
- Non-toxic: Refrigerants must not release toxic substances upon exposure to the environment causing harm to living beings or even non-living materials. This is the reason some refrigerant has been banned.
- Freezing point and boiling point: Boiling of refrigerant should be appreciably low i.e., below 0°C for smooth functioning inside the system. Also freezing point should be below operating temperature to prevent water clogging.
- Easily detectable in case of leakage: Refrigerants leakage can be detected by odor while some are odorless which in turn can contaminate the food products inside. To avoid leakage the working components are modified providing with a seal [15].
- Relatively inexpensive to produce: Refrigerants should be inexpensive as well as easily available.
- Low impacts on environment: Refrigerants have potential to deplete the ozone layer causing global warming. So selection should be such that it causes negligible impact on environment.

Pressure-Enthalpy (PH) Diagrams for Vapor Compression Cycle

Let us consider the process in the refrigeration cycle, where saturated liquid goes inside the expansion valve and comes out as saturated gas from it. In Figure 2, location (1) depicts the state of the refrigerant when it enters the compressor. The vaporized refrigerant at this point is at pressure, P_1 and enthalpy, h_1.

During the compression stroke, the vapor is compressed iso-entropically to pressure P_2, thus resulting in superheated vapor state shown in location (2). In location (2), the enthalpy has increased from h_1 to h_2. After that, the super heated refrigerant enters the condenser where the heat is removed and condensation occurs resulting in decrease of enthalpy from h_2 to h_4 as shown in location (3). However, the pressure remains constant, P_2 in this location. Finally, the pressure decrease to P_1 in the expansion valve but the enthalpy remains constant at h_4 as shown in location (4). Further, in location (4) it can be inferred that the refrigerant is in vapor as well as in liquid state. The mixture of vapor and liquid goes inside the evaporator and gets converted to vapor completely after accepting heat from substance to be cooled. Here pressure remains constant at P_1 and enthalpy changes from h_4 to h_1 [16].

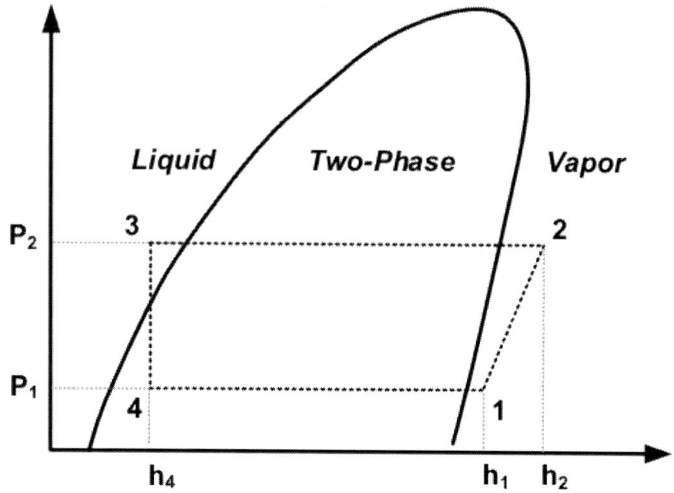

Figure 2. Pressure enthalpy diagram.

Applying mass and energy rate balances, we obtain the following equations:

- Evaporator: $Q_{in} = m\ (h_1 - h_4)$
- Compressor: $W = m\ (h_2 - h_1)$
- Condenser: $Q_{out} = m\ (h_2 - h_3)$
- Expansion Valve: $h_4 = h_3$ i.e., enthalpy remains same during throttling.

Coefficient of Performance of Refrigerator

It is very essential to master the measure of efficiency of a system. The Coefficient of Performance (CP) is the parameter used for evaluating the efficiency of a refrigeration system. Figure 3 demonstrates how the CP of VCRS is obtained.

$$CP = \frac{\text{heat captured by refrigerant in evaporator}}{\text{work done to the compressor}}$$

$$CP = \frac{h_1 - h_4}{h_2 - h_1}$$

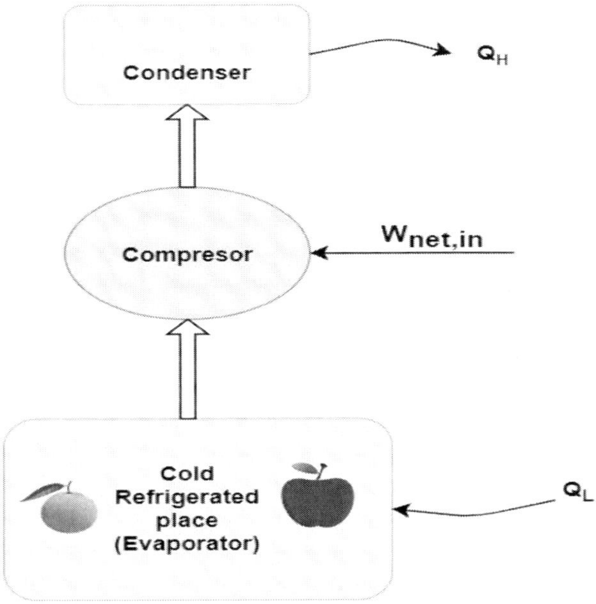

Figure 3. Coefficient of Performance (CP) diagram of VCRS.

Temperature-Entropy (TS) Diagrams for Vapor Compression Cycle

In Figure 4, the VCRS is plotted on temperature-entropy coordinates. Process 4 to 1 is compression and 2 to 3 is expansion which are isoentropic process represented by vertical lines. Process 3 to 4 is evaporation and 1 to 2 is condensation which is isothermal process represented by horizontal lines. The refrigerant is compressed into vapor inside the compressor. The vapor in this region is superheated. The superheated refrigerant then moves to the condenser rejecting its heat to the surrounding. Thus the superheated vapor turns into liquid. The liquid refrigerant flashes into liquid + vapor when passing through the expansion valve. The liquid + vapor refrigerant is converted to complete vapor inside the evaporator by acquiring latent heat from the substance to be cooled. The refrigerant then moves to the compressor and the cycle continues [17].

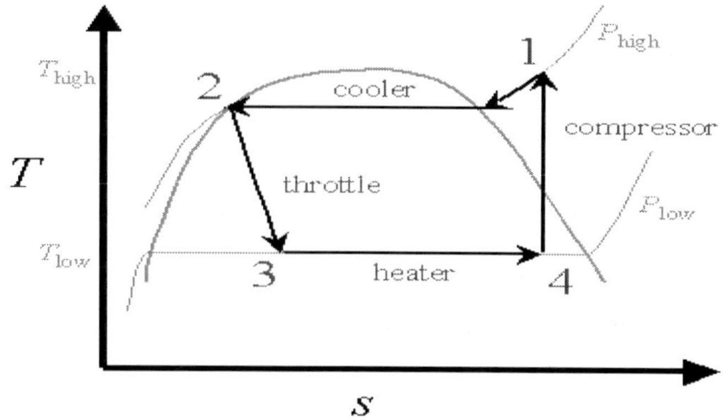

Figure 4. Temperature Entropy Diagram.

Actual Vapour Compression Cycle

The vapor compression cycle discussed above is theoretical vapor compression cycle but in practice the cycle used for cooling of substances is actual vapor compression cycle. It has less CP and requires more work. This cycle operates on the basis of Evans-Perkins cycle, also called as reverse Rankine cycle. In theoretical compression cycle there is no mention about fluid and mechanical friction. Friction generates heat which no longer makes the process isoentropic. This difference is explained in Figure 5.

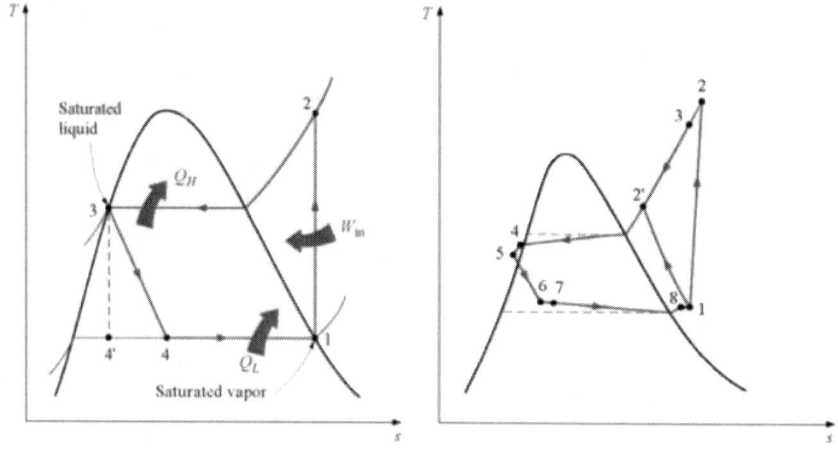

Figure 5. Temperature Entropy Diagram.

Evaporation

At point 1, In ideal VCRS refrigerant enters the compressor in saturated vapor form but in actual VCRS the refrigerant enters in slightly superheated form. The line connecting evaporator to compressor (8-1) signifies pressure drop as well as heat transfer to the surrounding.

Compression

In ideal VCRS refrigerant is compressed isoentropically meaning there is no change in entropy but in actual compressor is not internally reversible and hence entropy increases (4-5).

Condensation

Heat rejection and addition in condenser do not occur at constant temperature and pressure as a result of pressure drop in refrigerant.

Expansion Valve

Here ideal VCRS and actual VCRS differ in enthalpy. In actual process, enthalpy is lower compared to that of ideal.

Multi-Stage Refrigeration System

Single-stage system cannot be used when ultra-low temperature is desired [18]. It is often found that performance of single stage systems is sufficient as long as the difference in temperature between evaporator and condenser (temperature lift) is small. Up-liftment of temperature results in inefficiency of the single stage system. In such case there is need of another system where the compression ratio is too high to reach the temperatures necessary for evaporation and condensation of the steam [19]. Multi-staging is done to reduce power consumption [20]. Multi-stage refrigeration systems are classified into:

- Multistage Compression Refrigeration System
- Multistage Evaporator Refrigeration System
- Cascade Refrigeration System

Multistage Compression System

Multi compressor refrigeration system consists of two compressors, one low temperature evaporator and one high temperature condenser. As shown in Figure 6, there is a flash chamber where the refrigerant is segregated into saturated liquid and vapor. The saturated liquid from the flash chamber first passes between the expansion valve and then between the evaporator. Inside the evaporator latent heat is absorbed exchanging heat between the refrigerant and substance to be cooled. Then the refrigerant passes through the compressor and becomes superheated. It is then fed to the mixing chamber. The saturated vapor from the flash chamber directly passes to the mixing chamber. In the mixing chamber both (saturated vapor and superheated refrigerant) mixes and moves to the high pressure compressor.

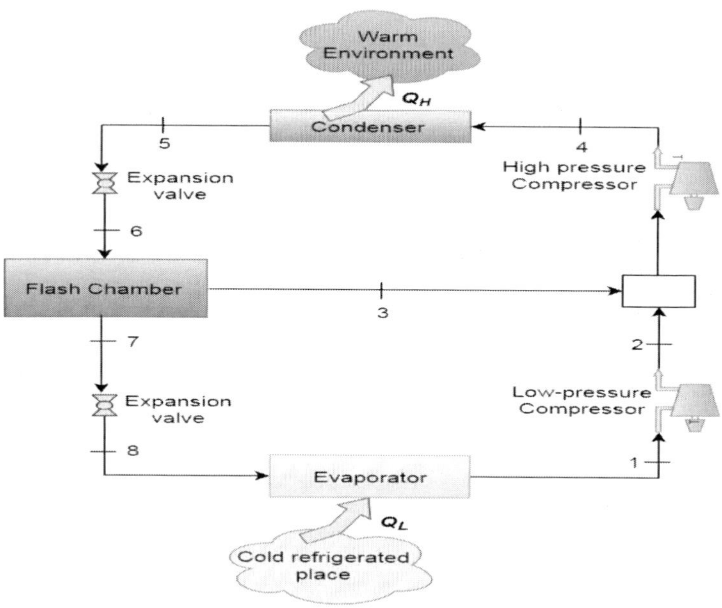

Figure 6. Multi-stage Compression Refrigeration System.

Multistage Evaporator System

Multi evaporator refrigeration system with individual expansion valve is shown in Figure 7. It consists of a condenser, a compressor, two or more evaporators. Each evaporator is connected to an expansion valve. The vapor from the compressor condenses in the condenser. Vapor from the condenser goes first to the evaporator 1 and then to evaporator 2 after getting expanded in the expansion valve connected to each evaporator. At the opening of evaporator 1 and 2, the pressure of the refrigerant are not same because expansion at evaporator 1 is earlier and at higher pressure. So it is connected to another expansion valve to expand the vapor at same pressure as evaporator 2 and mixing takes place. Then the mixture goes to the compressor and the cycle continues.

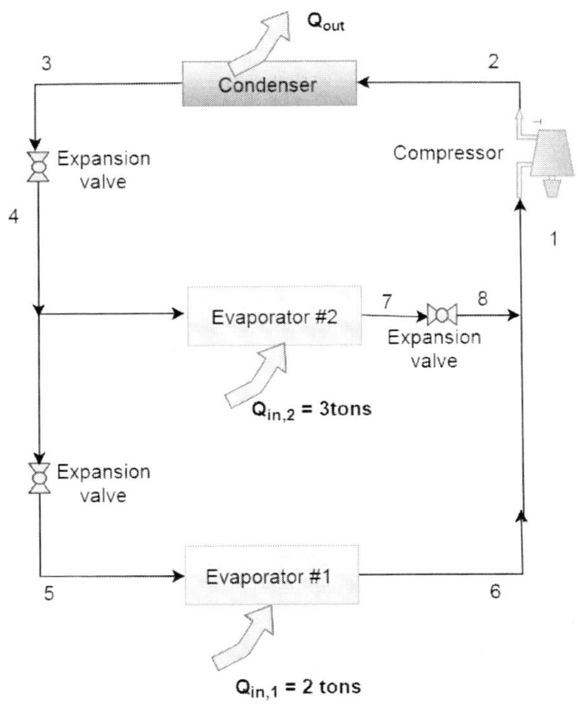

Figure 7. Multi-stage Evaporator Refrigeration System.

Cascade Refrigeration System

In order to achieve very low temperatures two or more vapor-compression systems can be connected in series, called *cascading* [21]. Two variants of refrigerants with two different boiling points are mostly utilized in cascade refrigeration system to achieve this very low temperature. Independent freezing cycle is carried out for each refrigerant but runs in proximity using the heat exchanger. The illustrative view of this system is shown below. This system is employed to obtain ultra-low temperatures ranging from -40 to -80°C.

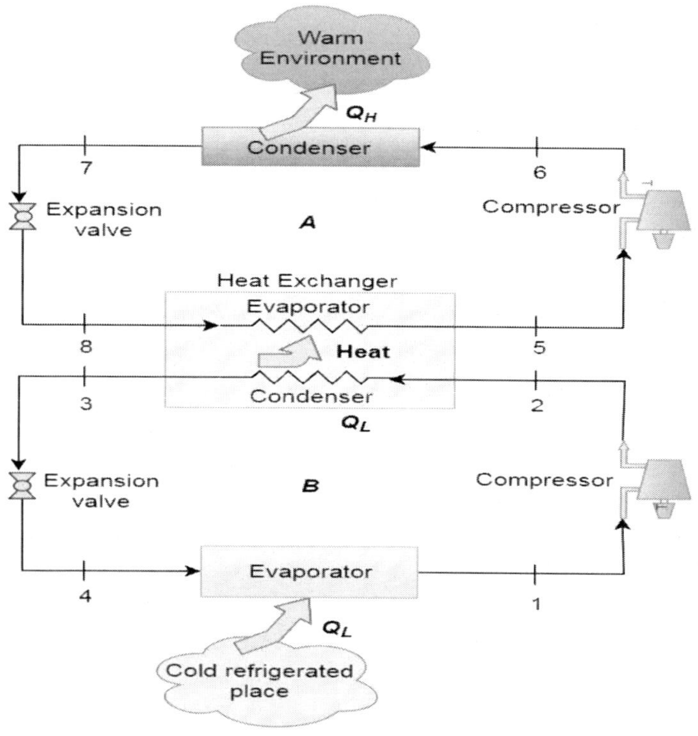

Figure 8. Cascade Refrigeration System.

Let us try to understand cascade refrigeration system from Figure 8. It consists of two vapor compression cycles A and B, where cycle B and cycle A are low temperature vapor compression cycle (LVCS) and high temperature vapor compression cycle (HVCS) respectively. The cascading is achieved using the heat exchanger where the evaporator and condenser of cycle A and

B are connected together. In cycle B the superheated refrigerant moves to the condenser from the compressor undergoing condensation and losing heat. The heat released from condenser is quickly absorbed by the evaporator in the heat exchanger. The heat exchange takes place because the temperature of the refrigerant at cycle B is lower than that of cycle A. Thus by using suitable refrigerants at the higher and lower temperature and pressure ranges, maximum benefit is derived.

Vapor Absorption Refrigeration System (VARS)

The six principal components of a VARS are evaporator, condenser, expansion valve, vapor generator, absorber and pump. Unlike vapor compression system, vapor absorption system consists of vapor generator, absorber and pump instead of compressor. This is the main difference between VCRS and VARS.

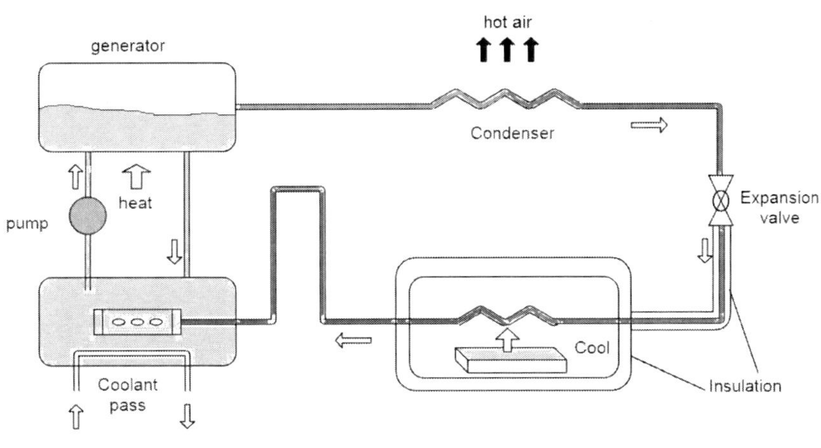

Figure 9. Vapor Absorption Refrigeration System.

In Figure 9, evaporator is connected to the absorber with pipes between them. Absorber contains ammonia and water. In VARS ammonia and water acts as refrigerant and absorbent respectively. Water and ammonia is pumped to the vapor generator where high heat is supplied. Heat supplied causes ammonia to liberate out from the cold water escaping the vapor generator at high pressure. Water comes back to the absorber. High pressure ammonia reaches the condenser inside which it condenses to liquid by losing its latent heat. This liquid ammonia enters the expansion valve where it undergoes sudden expansion causing both temperature and pressure to reduce to very low. With lower temperature and pressure ammonia then enters the evaporator and is exposed to the substance to be cooled (e.g., food).

Table 6. Difference between vapor compression and vapor absorption refrigeration system

Vapor absorption	Vapor compression
Heat energy is utilized	Mechanical energy is utilized
Wear and tear is less because of less moving parts	More wear and tear because of more moving parts
COP is less	COP is more
Charging of refrigerant is simple	Charging of refrigerant is difficult
Refrigerant used: ammonia, water or lithium bromide	Refrigerant used: ammonia, CFCs, HFCs etc.

Liquid ammonia extracts heat from the food converting itself to vapor of low pressure. This converted low pressure vapor reaches the absorber and the cycle continues. Table6 puts forward the major difference between vapor absorption and vapor compression refrigeration system.

Application of Refrigeration System in Food Industry

Cold Stores

A cold store is a large version of a refrigeration system. Cold stores play a very important role in bulk preservation of fruits and vegetables which are mostly perishable as well as seasonal. Hence it is an intermediate between producers and sellers.

Cold stores have all the components similar to a refrigeration system i.e., evaporator, compressor, condenser and expansion valve [22]. Cold stores are insulated containers and the insulating material mostly used is thermal or expanded polystyrene, cork, etc. At the recent times PUF panels are mostly used for insulation. Materials to be used depend on its applications. Insulating materials should have very low thermal conductivity. It should be weightless, odorless and low cost. Temperature of cold store is usually maintained at below 4°C.The perishables are kept in trays and cold air is circulated throughout the trays to keep it cool for longer period. This cold air that is circulated is formed by the process of refrigeration. Vapor compression systems are commonly used for the purpose. Evaporators are provided mostly at the top of the cold store chambers. Cold air is circulated through coils with the help of fans. The temperature between evaporating air and inlet air temperature is kept low to maintain high humidity inside the chamber. Humidity mostly preferred is 95%-98% depending on the products, those are kept inside. The next thing that should be taken care of is to maintain concentration of gases inside the store. CO_2 and O_2 levels should be maintained as recommended by commodity storage manual of World Food Logistics Organization. Loading, unloading and stacking of commodities is done in pallets, boxes, crates, bin, etc. Material handling equipments such as small hand operated trucks, pallet jacks, etc. are used inside the store. Loading and unloading is done with the help of fork lift truck and in case of meat it is hung by hooks. Refrigerant mostly preferred in cold stores is ammonia [23]. The other three components compressor, condenser and expansion valve are placed outside the cold store.

Refrigerated Tanks

Cooling tanks are mostly used in – dairy industry, beverage industry and in some cold supply chains. Purpose of a cooling tank is to absorb the heat from the contents inside the tank keeping it cool to a specific temperature and time. Design of a cooling tank is such that there is a large tank where the product is kept, a compressor outside, evaporator inside and condenser coils outside and operates in the similar way as vapor compression system. In industries, cooling tanks are mostly used for keeping milk around 4°C for a certain period of time and in the beverage industries it is used for chilling purpose. This chilling is important to keep beverages away from micro organisms to maintain its shelf life. Beer and wine making has maximum application of chilling tanks. Chilling is also important to remove heat that the product gains during mixing, pasteurization etc.

Industrial Freezers

Industrial freezers are mostly used for commercial purpose such as in industries related to bakery, meat & meat products, milk and milk products etc. operated by mechanical refrigeration [24]. It also works by the principle of Vapor Compression System having four main components and maintaining temperature below -18°C. Multi-stage systems are mostly used in industrial applications where refrigeration at different and sometimes very low temperatures is required. For example, in a dairy plant, for making ice cream, refrigeration temperature of –30°C is desired and for chilling milk, a temperature of almost 2°C is desired. In such cases use a multi-evaporator system may be recommended which operates at ultra low temperatures.

Refrigerated Chambers

Specialized stores or some advanced specialty whole sale super markets are equipped with refrigerated chambers for perishables. Safety is essential from field to fork. So it has become very essential to preserve the raw as well the processed products. This is the reason such stores consists of refrigerated chambers for agricultural produce as well as the processed bakery items, milk products, meat products, beverages and so on. In some cases the inside atmosphere of refrigerated chambers are controlled or modified with gases for better storage.

Refrigerated Vehicles

Simple *insulated* and *ventilated* vans are those which are fixed with cooling equipments and are mostly utilized for carrying fruits including some other fresh produce. Refrigerator trucks can be ice-cooled, implemented with any mechanical refrigeration systems or utilized carbon dioxide (either as dry ice or in liquid form) as a chilling agent. Unlike other refrigeration units, refrigerated trucks also consist of condenser, compressor, evaporator and throttling device. The functioning of these components is dependent on power driven from the vehicles engine. These are mostly used for transportation of ice-cream, meat, poultry, seafood etc. to long distance often for military services. Generally, the temperature range is between 0°C to -18°C.

Conclusion

Creating a very low temperature compared to its surrounding is the main purpose of refrigeration. In order to produce this challenging cooling effect, use of four basic components, viz- evaporator, compressor, condenser and expansion valve are employed appropriately. This chapter focuses refrigeration as an important preservation technique for upholding of food products both in household as well as in industrial purpose. With the evolvement of the process of refrigeration, it has occupied more space in the life of human beings. The technique of mechanical refrigeration was a breakthrough during the 19th century before which preservation of eatables was managed in natural ways. Mechanical refrigeration functions in a sequence of consecutive processes that form a cycle that give back the refrigeration effect. So familiarity with some basic terms like enthalpy, entropy, specific heat, latent heat and sensible heat is necessary. Based on its working mechanical refrigeration is of two types-vapor compression and vapor absorption refrigeration system. The vapor compression system operates in a cyclic fashion in which the working fluid called refrigerant passes through the four basic components-evaporator, compressor, condenser and expansion valve arranged in a sequence. Refrigerant flows from one component to another changing its state and thus produces refrigeration effect. Vapor compression system is also demonstrated by pressure-enthalpy and temperature-entropy diagrams. The working of vapor compression is more emphasized in this chapter compared to the other systems. The other systems which are described briefly are multistage refrigeration system and vapor absorption system. Multi stage refrigeration systems are used where ultra low temperature is needed. Industrial applications found its ways by making use of multistage refrigeration systems. Compared to single stage refrigeration system multi stage are more efficient and consumes less power to produce cooling effect. Single stage refrigeration systems may behave inefficiently when the temperature lift increases so multistage systems consisting of two or more compressors connected in series are used. Desired temperature for single stage refrigeration is +10°C to -30°C whereas for temperature below -30°C for two stage and for temperature below -60°C three stage is used. For this ability of ultra cooling effect, multistage systems have numerous industrial applications. The next is the cascade system where two different refrigerants thermally coupled are used. Cascade systems can be used in dry ice manufacturing, deep freezing etc. Unlike vapor compression system, vapor absorption system consists of vapor generator, absorber and pump instead of compressor, which is the main difference between the two. Refrigeration thus has tremendous use in food industries right from its use in preserving eatables in stores to its use in cold store where agricultural produce are stored to ultra low temperature.

Acknowledgments

Author is thankful to the entire fraternity of University of Science and Technology Meghalaya, Meghalaya, India for their immense help and support.

Declaration

This chapter is a comprehensive compilation of works done by various scientists, engineers and authors for academic purpose. The authors of this chapter took extensive care to cite most of the relevant previous authors in this chapter. If any name is missed in citation unknowingly, the authors also offer apology to them and honor their works also.

References

[1] Rech, S.; Finco, E.; Lazzaretto, A. A Multi-criteria Approach to Choose the Best Renewable Refrigeration System for Food Preservation. *Renewable Energy,* 2020, 154, 368-384.

[2] Heldman, D. R.; Hartel, R. W. Refrigerated storage. In: *Principles of Food Processing*. Springer, 1999, 83-112.

[3] Evans, J. Emerging refrigeration and freezing technologies for food preservation. *Innovation and future trends in food manufacturing and supply chain technologies*. Woodhead Publishing, 2016, 175-201.

[4] Kalla, S. K.; Usmani, J. A. Refrigeration cycles and systems: a review. *International Journal of Engineering Research and Science & Technology,* 2014, 3, 148-162.

[5] Popa, E. E.; Mitelut, A.; Popa, M. E. Trends in refrigeration technologies used for food preservation–a review. *Scientific Bulletin. Series F. Biotechnologies*, 2019, Vol. XXIII.

[6] Stoecker, W. F. *Industrial refrigeration handbook*. McGraw-Hill Education, 1998.

[7] Badr, O.; O'Callaghan, P. W.; Probert, S. D. Vapour-compression refrigeration systems. *Applied Energy,* 1990, 36, 303-331.

[8] Al-Rashed, Abdullah A. A. A. Effect of evaporator temperature on vapor compression refrigeration system. *Alexandria Engineering Journal,* 2011, 50, 283-290.

[9] Cofré-Toledo, J.; Vascoa, D. A.; Isaza-Roldán, C. A; Tangarife, J. A. Evaluation of an integrated household refrigerator evaporator with two

eutectic phase-change materials. *International Journal of Refrigeration,* 2018, 93, 29-37.

[10] Woo, S.; O'Neal, D. L.; Kim, Y. Reliability design and case study of a refrigerator compressor subjected to repetitive loads. *International Journal of Refrigeration,* 2009, 32, 478-486.

[11] Sahay, K. M.; Singh, K. K. *Unit Operations of Agricultural Processing*. 2nd Edition; Vikas Publishing House Pvt. Ltd; 1996; pages 386.

[12] James, S. J.; James, C. The food cold-chain and climate change. *Food Research International,* 2010, 43, 1944-1956.

[13] Riffat, S. B.; Afonso, C. F.; Oliveira, A. C.; Reay, D. A. Natural refrigerants for refrigeration and air-conditioning systems. *Applied Thermal Engineering,* 1997, 17, 33-42.

[14] Hundy, G. F. *Refrigeration, air conditioning and heat pumps*. 5th Edition, Butterworth-Heinemann, 2016.

[15] Fukuta, M.; Takeru S.; Masaaki M. Leakage and friction characteristics at sliding surface of tip seal in scroll compressors. *International Journal of Refrigeration,* 2021, 125, 104-112.

[16] Fellows, P. J. *Food Processing Technology: Principles and Practice*. 3rd Edition; Elsevier; 2009; pages 913.

[17] Singh, R. P.; Heldman, D. R. *Introduction to food engineering*. 4th Edition; Elsevier Academic Press; 2008; pages 864.

[18] Jung, D. S.; Radermacher, R. Performance simulation of single-evaporator domestic refrigerators charged with pure and mixed refrigerants. *International Journal of Refrigeration,* 1991, 14, 223-232.

[19] Baakeem, S. S.; Orfi, J.; Alabdulkarem, A. Optimization of a multistage vapor-compression refrigeration system for various refrigerants. *Applied Thermal Engineering,* 2018, 136, 84-96.

[20] Rajput, R. K. *Refrigeration and Air-Conditioning*. S K Kataria & sons, New Delhi; 2009; pages 900.

[21] Pan, M.; Zhao, H.; Liang, D.; Zhu, Y.; Liang, Y.; Bao, G. A review of the cascade refrigeration system. *Energies,* 2020, 13, 2254.

[22] Moerman, F.; Fikiin, K.; Evans, J.; Foster, A. Air infiltration control to reduce hygiene hazard in refrigerated food processing and storage facilities. In: *Hygienic Design of Food Factories*. 2nd Edition; Woodhead Publishing, 2020.

[23] Krysanov, K. S., Belukov, S. V., Pribytkovskiy, N. V., & Dibirasulaev, D. M. (2020). Efficiency of Cold Generation Cycles using Various Refrigerants. *Chemical and Petroleum Engineering*, *56*(7), 646-653.

[24] Tassou, S. A.; Lewis, J. S.; Ge, Y. T.; Hadaweya, A.; Chaer, I. A review of emerging technologies for food refrigeration applications. *Applied Thermal Engineering*, 2010, 30, 263-276.

Chapter 6

Therapeutic Potential Applications for Bee Pollen

Sofia Guiné Florença[1,*], Ofélia Anjos[2,3,4], Manuela Ferreira[5], Maria Graça Campos[6], Letícia M. Estevinho[7], Paula Correia[8], Cristina Amaro da Costa[8] and Raquel Guiné[8]

[1]FCNAUP, University of Porto, Porto, Portugal
[2]School of Agriculture, Polytechnic Institute of Castelo Branco, Castelo Branco, Portugal
[3]Forest Research Centre, School of Agriculture, University of Lisbon, Lisbon, Portugal
[4]CBP-BI, Castelo Branco, Castelo Branco, Portugal
[5]UICISA:E Research Centre and School of Health, Polytechnic Institute of Viseu, Viseu, Portugal
[6]Coimbra Chemistry Centre (CQC, FCT Unit 313) (FCTUC) and Observatory of Drug-herb Interactions, University of Coimbra, Coimbra, Portugal
[7]Centro de Investigacao de Montanha (CIMO), Instituto Politécnico de Bragança, Bragança, Portugal
[8]CERNAS Research Centre, Polytechnic Institute of Viseu, Viseu, Portugal

Abstract

Since ancient times, bee pollen has emerged as a rich food product, which, besides being a source of energy, contains beneficial substances with recognized properties, both nutritionally and therapeutically. Recently, research about this natural product has allowed a more

---

* Corresponding Author's E-mail: sofiaflorenca@outlook.com.

In: Food Processing
Editors: Myriam Huijs
ISBN: 978-1-68507-570-5

supported recognition of its health enhancement potential. The therapeutic properties of bee pollen are associated with the chemical composition, in particular vitamins, minerals, polysaccharides, and bioactive compounds, like phenolic acids and flavonoids. Due to the reasons highlighted, this review aims to synthesize some fundamental aspects of the relationship between the product's constituents and their respective bioactivity, in light of the most recent publications, which reveal the potential of bee pollen as a healthy product. Among them, citations of the most relevant benefits report the antioxidant, anti-inflammatory, antifungal, and antimicrobial bioactivities. Other potentialities, as the therapeutic use in drug discovery, include, for instance, the ability to modulate cancer development. Along the sections of this chapter these health effects will be presented and discussed.

Keywords: bee pollen, chemical composition, bioactive compounds, antioxidant activity, anticancer effect

Introduction

By general definition, pollen is a coarse powdery substance composed of male gametes of seed plant flowers, and whose main function is reproduction. However, the product "Bee pollen (BP)" that is used for nutrition purposes is a mixture of various kinds of pollens from different flower species (Figure 1), which include enzymes produced by bees, such as amylases and catalases. These agglomerates vary in size, shape, color, characteristics, and flavour [1].

Figure 1. Samples of bee pollen from different flower species.

Since ancient times, this product has been used in traditional medicine to improve a wide number of diseases or acute conditions, such as colds, flu, ulcers, anemias, colitis, allergies, among others [2,3]. Additionally, it has also been reported as being used to fight the effect of premature ageing [4], maybe due to antioxidant activity attributed to some floral sources.

Recently, several food and nutrition advisors have suggested daily supplementation with BP. The main allegations are based on the numerous health benefits that have been claimed [5–7]. The main advantages described are: being an immune system and tissues' resistance stimulator [6–8], being beneficial to the skin [7,8], respiratory system [7,8], digestive system [6–8], cardiovascular system [7,8], fertility [7,8] and in addiction treatments [6–8]. Additionally, it may act as an antiallergic agent when its exposure is tightly controlled [6,7]. These benefits come from the numerous constituents that this product contains, namely, vitamins, minerals, amino acids, phenolic acids, and flavonoids. This last, for instance, has been studied for various biologic properties as antioxidant and anti-inflammatory, anticancer, antibacterial, antifungal, hepatoprotective, anti-atherosclerotic, antiallergic, and capable of modifying and regulating the immune system [6–8] [2,3,9].

Nutritional Composition

The nutritional composition of BP is highly variable, as it depends on geographic origin, bee species, ecological habitat, and even the season of the year [2,9]. In its composition, there are about 250 substances, including carbohydrates (40-85%), proteins (14-30%), with a total of 20 amino acids, lipids (1-10%), vitamins, minerals, polyphenols, carotenoids and flavonoids (3-8%) [3,10–12].

Table 1 show the main macro components found in BP. Among them, carbohydrates are the major components, varying from 130 to 550 mg/g in pollen, followed by protein content, also present in substantial amounts (varying from 100 to 400 mg/g).

Table 2 presents the main micro components in bee pollen. The primary vitamins are vitamin C, A, E, B_1, B_2, B_3, B_6, folic acid, and biotin. With regard to minerals, pollen contains potassium, phosphorus, calcium, magnesium, zinc, manganese, copper, and iron [2,9,10].

Table 1. Reported macro components of bee pollen and nutritional requirements

Component	Amount (g/100 g) [13]	RDI[1] (g/day) [14]
Total carbohydrates (mainly sacharaides)	13–55	130
Dietary fiber, pectin	0.3–20	38
Protein	10–40	46-56
Fat	1–13	--
Ash	2–6	--

[1]Required Daily Intake.

Table 2. Reported micro components of bee pollen and nutritional requirements

Component	Amount (mg/100 g) [13]	RDI[1] (mg/day) [15–17]
Vitamins		
Beta-carotene (provitamin A)	8.4–10.8	0.9
Thiamine (vitamin B_1)	1.1–1.2	1.1
Riboflavin (vitamin B_2)	1.1–1.3	1.3
Niacin (vitamin B_3)	14–16	15
Pantothenic acid (vitamin B_5)	5	6
Pyridoxin (vitamin B_6)	1.3	1.4
Folic acid (folate, vitamin B_9)	0.4	0.4
Ascorbic acid (vitamin C)	75–90	100
Biotin (vitamin H)	0.03	0.045
Tocopherol (vitamin E)	15	13
Minerals		
Potassium (K)	2.6–3.4	2000
Phosphorus (P)	700	1000
Calcium (Ca)	1000	1100
Magnesium (Mg)	310–420	350
Zinc (Zn)	3–25	8–11
Manganese (Mn)	2–11	1.8–2.3
Iron (Fe)	1.1–17	18–8
Copper (Cu)	0.2–1.6	0.9

[1]Required Daily Intake.

Bioactive Components ahd Health Effects

Among bee products, BP has demonstrated various nutritional benefits and bioactive properties, which allow classifying them as excellent functional foods or superfoods [18]. Mărgăoan et al. [15] conducted a recent review on the therapeutic properties of different flora sources of pollens, highlighting the studies that confirm its antioxidant, antimicrobial, anti-carcinogenic, anti-inflammatory anti-osteoporosis, and hepatoprotective activities. Also, research has been conducted about the potential of certain compounds from pollen which can be explored as future adjuvants for oncologic therapy.

Other important beneficial health effects of BP are related to vitamins and minerals that have relevant functions on the human body. Further compounds found in this product are phenolic acids and flavonoids. These chemical structures include aromatic ring(s) linked to one or more hydroxyl groups. Due to these characteristics they have been reported as having high antioxidant capacity, thus protecting the cells against oxidative stress [19,20], [21–25]. In addition, the polysaccharides have been reported as possible allies for antineoplastic treatments [26,27].

Antioxidant Activity

The antioxidant effect of BP is attributed to various groups of compounds present, such as phenolic compounds cited above, but also to carotenoids, vitamin C and E and glutathione, among others. These substances have the ability to remove, directly or indirectly, free radicals and toxic metals, and in a certain way, be involved in processes of protection against these mutagenic agents, acting as free radical scavengers and lipid peroxidation inhibitors [2,9,28] [29].

Various publications reported data about the bioactivity of BP. For instance, Anjos et al. [30] evaluated the total phenolic compounds and flavonoids content of BP, as well as the phenolic profile and the antioxidant bioactivity of its extracts. They reported a higher activity of the floral sources under study when compared to other sources found in studies published by other authors [31,32]. The work by Almaraz-Abarca et al. [33] also evaluated the antioxidant activity of a monofloral BP using an *in vitro* model (inhibition of lipid peroxidation on mouse hepatic microsomal preparations), and also an *in vivo* system (homogenized liver of bromobenzene-intoxicated mice) by quantification of thiobarbituric acid-reactive substances (TBARS). They concluded that BP from the floral samples studied is an important source of flavonoids, with natural antioxidant activity. The work performed by Mărghitaş et al. [34] reported a great variability in the antioxidant activity and total phenols' content of BP originating from different botanical species, which corroborates the data shown by the other authors. In fact, Gonçalves et al. [35], applying fuzzy classification model could clear identify the influence of botanical origin of BP in their chemical composition and subsequent antioxidant activity. Aylanc et al. [36] completed this information with an interesting investigation where they evaluated the bioaccessibility and antioxidant activity of BP by simulating the gastrointestinal digestion and reported just a slight decrease in the level of total phenolics and flavonoids along the gastrointestinal tract.

Anti-Inflammatory Activity

The anti-inflammatory effect is also linked to the previous bioactivity and the polyphenols, including flavonoids. They can induce a beneficial effect on several cells, which play a fundamental role in the body's defense against pathogens and inflammatory processes [2,3,9].

Maruyama et al. [37] evaluated the anti-inflammatory effect of ethanolic extracts from Spanish BP and concluded that the bee pollen showed a high anti-inflammatory capacity, partially attributed to some flavonoids present. The authors hence suggest that BP could be helpful as a dietary supplement, as well as a functional food.

The work performed by Pascoal et al. [38] assessed the anti-inflammatory activity of eight BP samples purchased in the Portuguese market. The authors reported that all samples demonstrated anti-inflammatory bioactivity, with percentages of inhibition varying from 12 to 25%. In line with this research, Feás et al. [39] analysed twenty two samples of Portuguese organic BP, and found percentages of inhibition in the same range, up to 25%.

The data found in literature gives good hope of potential activity of BP. However, additional research is still necessary to validate this evidence of efficacy and safety in the most promising flower sources.

Antimicrobial and Antifungal Activity

The antibacterial and antifungal effect is associated with the existence of an oxidase produced by bees and phenolic acids and flavonoids that manage to degrade the cytoplasmic membrane of some species of bacteria and fungi [2,3,9].

Ozcan [40] investigated the inhibitory effects on the growth of *Aspergillus parasiticus* strain NRRL 2998 *in vitro* produced by pollen and propolis extracts prepared with samples collected in different parts of Turkey. The results showed that 2% of the pollen extract was the less active concentration targeted for the fungus tested.

Morais et al. [43] tested the antimicrobial activity of five bee pollens originating from Portuguese Natural Parks, using eight microorganisms: *Bacillus cereus* ESA 55, *Escherichia coli* ESA 15, *Staphylococcus aureus* ESA 72, *Salmonella typhi* ESA 101, *Zygosacharomyces mellis* ESA 35, *Zygosacharomyces bailii* ESA 1307, *Zygosacharomyces rouxii* ESA 23, *Candida magnólia* ESA 11). Their results demonstrated antimicrobial activity of all the tested extracts, although with differences in terms of selectivity and Minimum Inhibitory Concentration for each microorganism. They also observed that the type of pollen influenced the growth of the microorganisms under study.

Related investigation was published by Feás et al. [39]. The authors also evaluated the antimicrobial activity of Portuguese BP samples against seven microorganisms: *Staphylococcus aureus* ATCC 6538, *Staphylococcus aureus* ESA 159, *Pseudomonas aeruginosa* ATC, *Pseudomonas aeruginosa* ESA 22, *Escherichia coli* ATCC, *Escherichia coli* ESA 37, and *Candida glabrata* ATC). The results showed that the antimicrobial activity differed significantly depending on the pollen sample, for all the microorganisms studied. The two strains of *Staphylococcus aureus* were the most sensitive, with minimum inhibitory concentrations ranging from about 2 to 4 mg/mL.

Pascoal et al. [38] provided data from the bioactivity of BP samples against seven microorganisms: *Staphylococcus aureus* ATCC 6538, *Staphylococcus aureus* ESA 159, *Pseudomonas aeruginosa* ATC, *Pseudomonas aeruginosa* ESA 22, *Escherichia coli* ATCC, *Escherichia coli* ESA 37, and *Candida glabrata* ATC. They again found that all samples exhibited antimicrobial activity, with *Staphylococcus aureus* being again the most sensitive and *Candida glabrata* the most resistant of all the microorganisms studied.

However, not all research studies were so positive in demonstrating the antimicrobial effect of BP. For example, in a study by Carpes et al. [42] the antimicrobial potential of 36ix pollen extracts from south Brazil was investigated using nine microorganisms: *Agrobacterium tumefaciens*, *Bacillus subitilis*ATCC 21.332, *Streptococcus mutans Ingbritt* 1600, *Staphylococcus aureus* ATCC 25.923, *Klebsiella pneumoniae*, *Xanthomonas vesicatoria*,

Xanthomonas axonopodis, Pseudomonas aeruginosa ATCC 15.442, and *Pseudomonas syringae*). Their results showed that despite the high concentrations of phenolic compounds and flavonoids in the pollen extracts, these couldn't inhibit the growth of the tested microorganisms.

Hepatoprotective Effect

The hepatoprotective potential of BP from the chestnut flower was assessed by Yıldız et al. [43], in an *in vivo* rat model. Their results demonstrated the effectiveness of BP to protect the hepatocytes from the oxidative stress, thus promoting the healing of the damage induced in the liver of the rats.

Later, Šarić et al. [44] investigated the in vivo oxidant and antioxidant effects of flavonoids from Croatian BP. The results revealed that the supplementation with this product, in a concentration of 100 mg/kg body weight, modulated antioxidant enzymes in the liver of mice and reduced hepatic lipid peroxidation.

Bevzo and Grygor'eva [45] studied the effect of BP extract on glutathione system activity in the liver of mice under X-ray irradiation. Their results suggested that small X-irradiation doses activate the lipid peroxidation and antioxidant system enzymes in the liver. However, the administration of BP extract to the mice allowed normalizing the activity of several glutathione system enzymes in the liver.

Eraslan et al. [46] investigated the *in vivo* effect of the pesticide carbaryl on some biochemical changes in rats and the ameliorative effect of BP. The results obtained confirmed that carbaryl produced negative changes in the majority of the oxidative stress markers, as well as in the biochemical parameters investigated in the serum. However, these effects were minimized with the administration of BP extracts in dosages of 50 and 100 mg/kg body weight/day.

Anti-Atherosclerotic and Anti-Diabetic Effect

The anti-atherosclerotic activity is, on one hand, related to the existence of hypolipidemic compounds, which reduce the total concentration of lipids, triacylglycerols, and cholesterol. On the other hand, it is also associated with the presence of free fatty acids that are important precursors of prostaglandins, the main inhibitors of platelet aggregation, having a protective effect against cardiovascular disease [2,3,9].

The protective effects of flavonoids from pollen (*Brassica campestris* L.) on tert-butyl hydroperoxide-induced injury in the EA.hy926 cell line were evaluated by Yang et al. [47]. Their results confirmed that the pollen flavonoids exhibited anti-inflammatory effects and allowed maintaining normal redox and mitochondrial functions, as well as lipid metabolism in cells.

In another study, Laaroussi et al. [48] evaluated the effect of BP extracts against D-glucose-induced type 2 diabetes in rats. The primary antioxidant compound present in these BP samples was apigenin. The diabetes induced in the rats was reversed by administering BP extracts, restoring the biochemical parameters, and attenuating the harmful effects of D-glucose on liver and kidney functions.

Despite these promissing previous data, more research should be done to assure the effectiveness and safety of BP for this purpose.

Anti-Allergic Effect and Immune System Regulation

The antiallergic activity comes from the inhibition, by compounds present in pollen, of the secretion of histamine, the main responsible for the allergic response [2,3,9]. The bee pollen extracts contain phenolic compounds, particularly flavonoids, which have demonstrated immunomodulatory activity. In the research developed by Medeiros et al. [49] was evaluated the effect of BP extracts and the flavonoid myricetin in the model of ovalbumin-induced allergy. The treatment with the extract in dosages of 200 mg/kg showed inhibition of the paw oedema, and the treatment with the isolated myricetin at 5 mg/kg also inhibited pulmonary cell migration and IgE and IgG1 OVA-specific production. These results allowed the researchers to conclude that myricetin has antiallergic effect, which makes it a possible helper in the treatment of allergies.

The microbiota is crucial for the regulation of the immune system. Some compounds are able to contribute to this microbial diversity in the gut, for instance polysaccharides, flavonoids and sterols which can be also found in BP [2,3]. Zhao et al. [50] evaluated the effect of polysaccharides extracted from wolfberry Chinese BP on the immune response and gut microbiota composition in cyclophosphamide-treated mice. The obtained results evidenced that those polysaccharides enhanced the thymus and spleen indices, while promoting the production of immune-related cytokines in the serum of the mice. Among other data, the authors verified that the BP polysaccharides in this study could increase the amounts of Family_XIII, Ruminococcaceae, and Saccharimonadaceae, while promoting a decrease of the relative amounts of Tannerellaceae, Marinifilaceae, and Rikenellaceae, resulting in the promotion of the immune function.

Anticancer Effect

Anticancer activity is, to a great extent, the consequence of the antioxidant and various anti-mutagenic properties [2,3,9], for instance, the potential to reduce angiogenesis, cell growth, and proliferation. Arung et al. [51] tested the effect of seven BP samples on some cancer lines (MCF-7, HeLa and Caco-2 cancer cell lines). The BP extracts from *Homotrigona fimbriat* displayed more cytotoxicity than extracts of *Tetragonula testaceitarsis*, T*etragonula laeviceps*, *Tetragonula sarawakensis*, *Heterotrigona bakeri*, *Heterotrigona itama*, and *Tetragonula fuscobalteata*. Additionally, the cytotoxicity of *H. fimbriata* was 82.2% in MCF-7, 55.5% in HeLa, and 61.4% in Caco-2 cells, which was higher than the control.

The work by Pascoal et al. [38] focused on the antimutagenic activity of eight commercial BP samples purchased in Portugal. Their results revealed that all samples possessed antimutagenic activity, although some of them were more effective in decreasing the number of gene conversion colonies and mutant colonies when compared to others. This highlights that the floral source can influence their therapeutic properties.

Croatian BP samples were investigated by Šarić et al. [44]. The authors studied the possible effects of the flavonoids in the extracts for the estrogenic/anti-estrogenic activity and gene expression profile, using an in vivo model. The supplementation with BP (100 mg/kg body weight) induced 25% of anti-estrogenic effect without estrogenic activity. Additionally, the authors found that this supplementation induced a differential gene expression profile analyses, after BP enriched diet, which affected the gene expression of liver and brain cell models. Those are important in the apoptosis pathway and chemotaxis. Hence, the authors suggest that BP constitutes a rich source of compounds with health-protective potential.

The genotoxicity and antigenotoxicity of BP samples and their extracts (from *Cystus incanus* and *Salix alba*) was investigated by Pinto et al. in yeast and human cells (lymphocytes) [52]. Their results showed that the pollens in the essay were not genotoxic and were able to reduce the chromosomic damages induced by the cancer drugs tested. In view of their findings, they suggest that BP from the floral sources used in this work could be used as a food supplement with potential as chemoprotective properties. However, safety evaluation should be done in further research studies.

Furusawa et al. [39] studied the antitumour potential of the water-soluble fraction of pollen extract on lung carcinoma in an *in vivo* model (mice), and they found that the extract was cytotoxic in cell cultures at concentrations up to 2.5 mg/mL. Additionally, when used at concentrations of 0.5 g/kg, it significantly extended the lifetime of mice, without any adverse symptoms. Finally, it was found a beneficial synergetic therapeutic effect when it was combined with standard antitumor drugs, thus allowing to prolong the mice lifetime.

Wan Omar et al. [53] evaluated the antiproliferative effect of methanolic extracts of BP from Malasya on breast MCF-7 cancer cells, and also the synergistic effect of the pollen extracts with a chemotherapeutic drug (cisplatin). Their results indicated that these extracts have the potential to act as chemopreventive agents and can be explored in further investigations as a complementary treatment along with chemotherapy drugs, thus allowing the possible reduction in the required dose of those chemotherapy agents.

Tuoheti et al. [54] used high hydrostatic pressure to further improve the bio-functionality of BP, thus increasing its anti-proliferative properties in prostate cancer PC-3 cells. The results confirmed that the high-pressure treatment in fact enhanced the anti-prostate cancer activity of Lotus species BP, by increasing the bioactive metabolites.

Food Safety Concerns

Natural products are commercialized as safe options for the prevention and treatment of several diseases of varied nature. However, it is essential to acknowledge that consumers are not always informed about the potential side effects or even health risks arising from their usage. This is far more concerning, having in mind that people tend to self-medicate with these products, which in turn may not have passed a quality control or clinical trials to examine their efficacy and safety [12].

To date, only a few countries have approved legislation to regulate BP such as, for instance, Poland, Bulgaria, Switzerland, Brazil and Argentina [11]. Nevertheless, having in mind this need, the Technical Committee for Food Products of the International Organization for Standardization established a subcommittee for beekeeping products (ISO/TC34/SC19). This technical group has recently begun standardizing the products, and in a near future, they will complete the necessary guidelines. Meanwhile, due to the lack of regulations, BP of different origins and floral sources can be characterized by heterogeneous nutritional properties. Besides that, it can have different levels of risk concerning food safety.

A recent review by Végh et al. [55] has focused on the food safety risks associated with the ingestion of BP, as well as the reported concentration of toxic substances present in the product. They postulate that these harmful substances may be due to contamination with pesticides or pesticide residues from the environment, but also other harmful substances may be found in this product. Among them stand heavy metals and metalloids. Molds can also be responsible for some contamination with mycotoxins.

Pesticides are used widely as essential allies for the food supply of the global population at present, which implies that residues are usually present in plant foods [56]. Although organic farming can contradict this tendency by producing foods in a more environmentally friendly way and with much less impact on the ecosystems, it is a fact that its expression at the global scale is still limited [57,58]. Pollinator protection relies on the determination of the bees' pesticide exposure, which in turn can be estimated based on pollen for being their main source of nutrients, apart from nectar. Many scientific works investigate the contents of pesticide residues in various pollen samples [59–64]. However, the reported results are highly variable, due to the influence of several factors such as the botanical origin, the method and time of application of the pesticides, the doses used, the physicochemical properties of the active substances, and environmental conditions like temperature, humidity, water activity and soil characteristics [59].

Several studies have proven that bees and beekeeping products can serve as biological indicators of the heavy metal contamination in a certain geographical area [65–69]. This means that bees are in deep contact with these compounds, which may be problematic as they consume the products which they stored in summer also during winter. In this way, they can be exposed to environmental contaminants all year round [70,71]. Maximum limits have been proposed by Campos et al. [13] for plumb (500 μg/kg), cadmium (30 μg/kg), and mercury (10 μg/kg).

Mycotoxins are secondary metabolites produced by filamentous fungi that evoke a toxic response to higher organisms, even when in low concentrations. They occur mainly in foods of plant origin, and the growth of mycotoxin-producing molds is dependent on several factors, being the most relevants temperature, relative humidity of the atmosphere and moisture content of the product, or oxygen concentration in the surrounding air [72]. Kostić et al. [73] suggest that BP composition constitutes a privileged spot for the development of several microorganisms, among which stand molds that produce mycotoxins. Although the European Commission has established maximum permissible amounts of some mycotoxins in several foods (Commission Regulation 1881/2006/EC), the beekeeping products were left out. Campos et al. [13] proposed limit values for aflatoxins back in 2008 (B1: max 2 μg/kg; B1+B2+G1+G2: max 4 μg/kg) and for the total amount of yeasts and molds (max 50,000 CFU/g).

Pollens derived from certain plant species contain natural hepatotoxic pyrrolizidine alkaloids, and these can be present in fairly high concentrations [74]. The pyrrolizidine alkaloids are secondary metabolites that contribute to the survival of certain plant species, acting as a part of their natural defense mechanism. More than 660 pyrrolizidine alkaloids have been identified so far [75], and their toxicity is variable according to their chemical structure [76]. Many studies are being carried out in order to establish the intervals for chronic and acute exposures to pollen-based products.

BP can also contain allergens (proteins) [77] and/or pollen grains from genetically modified plants [78], and these can have adverse effects when ingested.

Some possible common hazardous substances which can be present in bee pollen are:

- Pesticide active substances: chlorpyrifos, fluvalinate, carbendazim, thiacloprid [61,79];
- Toxicologically important elements: arsenic, cadmium, mercury, lead [13,80];
- Mycotoxins: aflatoxin-B1, ochratoxin-A, fumonisins, zearalenone, deoxynivalenol, T-2 toxin) [73,81];
- Pyrrolizidine alkaloids [74].

Végh et al. [55] concluded that pesticide residues do not seem to be a problematic risk for consumers, although they alert that the estimated acute exposure values can be close to the acute reference dose [55]. Concerning the heavy metals, they found that arsenic, cadmium, and lead can constitute a potential health risk to consumers, more than mercury. However, the studies with mercury are still scarce to make a definite conclusion [13,66,80,82,83]. Hence, it is advised to establish a maximum limit for this kind of substance to be present in BP and derived commercial products, as well as to effectively monitor their concentration in products aimed at human consumption [13]. Also, the pyrrolizidine alkaloid content of BP was highlighted as a risk factor for human health, and again its presence must be subject to control [84,85]. However, these pyrrolizidine alkaloids were found in a few plant species visited by bees.

Furthermore, the lack of available scientific data for the contamination of BP with mycotoxins and their amounts, drives the scientific community into the need of further studies on this area, in order to establish undoubtedly the problems related with the consumption of BP or its derived products [55].

Conclusion

The evidence available in the literature confirms that BP has numerous health-promoting properties, allowing it to be classified as a food with nutritional and therapeutic properties. Based on the current knowledge, it is possible to confirm most of the claims made for this product, namely that it has a beneficial effect on the respiratory system, due to its anti-inflammatory properties, on the cardiovascular system, on aging, due to antioxidant effect, and as an antiallergic agent. However, supplementation with BP is not yet fully supported by science, as its properties may vary significantly depending on floral sources or processing methodologies. Additionally some bee pollens can contain toxic and hazardous substances. Therefore, further analyses of risks versus benefits and limit doses are still needed.

Acknowledgments

The authors would like to acknowledge the following supporting organizations/projects:

- FCT - Foundation for Science and Technology, I.P., within the scope of the project Refª UIDB/00681/2020 (CERNAS research Centre) and project Refª UIDB/00239/2020 (Forest Research Centre).
- Project beeB - Foster for beekeeping bridges through innovative and participative training (ref. 2019-1-PT01-KA202-060782)
- Project "FZ – drone flour" (ref. PROJ/IPV/ID&I/013 in the ambit of partnership IPV-CGD (Caixa Geral de Depósitos).
- CERNAS Research Centre and the Polytechnic Institute of Viseu.

The authors thank the valuable advice of the experts who revised the present chapter: Professor Kemilla Rebelo, Institute of Health and Biotechnology, Federal University of Amazonas, Coari-AM, Brazil; Professor Wilma Spinosa, University of Londrina, Londrina-PR, Brazil.

References

[1] Campos MG, Anjos O, Chica M, Campoy P, Nozkova J, Almaraz-Abarca N, et al. Standard methods for pollen research. *Journal of Apicultural Research* 2021;60:1–109. ttps://doi.org/10.1080/00218839.2021.1948240.

[2] Denisow B, Denisow-Pietrzyk M. Biological and therapeutic properties of bee pollen: a review. *J Sci Food Agric* 2016;96:4303–9. https://doi.org/10.1002/jsfa.7729.

[3] Komosinska-Vassev K, Olczyk P, Kaźmierczak J, Mencner L, Olczyk K. Bee Pollen: Chemical Composition and Therapeutic Application. *Evid Based Complement Alternat Med* 2015;2015. https://doi.org/ 10.1155/2015/297425.

[4] Kozmus P, Noc B, Vrtacnik C. *No bees, no life.* 1st ed. Ljublijana, Slovenia: Beebooks; 2017.

[5] Anjos O, Estevinho ML, Campos MG. Poderá o pólen ser onsiderado um superalimento? *Revista Do Instituto Politécnico de Castelo Branco* 2017;11:20–6. [Can pollen be considered a superfood? *Castelo Branco Polytechnic Institute Magazine*]

[6] Coelho TR. Pólen de Abelha: Para que serve, o que é e benefícios? *Myprotein*; 2017. [Bee Pollen: What is it for, what is it and benefits?]

[7] Bosch L. 10 Ways Your Health Can Be Improved By Eating 1 Spoon of This Unappreciated Bee Product. *FoodMatters*; 2013.

[8] Khalifa SAM, Elashal MH, Yosri N, Du M, Musharraf SG, Nahar L, et al. Bee Pollen: Current Status and Therapeutic Potential. *Nutrients* 2021; 13:1876. https://doi.org/10.3390/nu13061876.
[9] Li Q-Q, Wang K, Marcucci MC, Sawaya ACHF, Hu L, Xue X-F, et al. Nutrient-rich bee pollen: A treasure trove of active natural metabolites. *Journal of Functional Foods* 2018;49:472–84. https://doi.org/10.1016/j.jff.2018.09.008.
[10] Campos M, Frigerio C, Lopes J, Bogdanov S. *What is the future of Bee-Pollen?* 2010. https://doi.org/10.3896/IBRA.4.02.4.01.
[11] Thakur M, Nanda V. Composition and functionality of bee pollen: A review. *Trends in Food Science & Technology* 2020;98:82–106. https://doi.org/10.1016/j.tifs.2020.02.001.
[12] Campos MG, Olena L, Anjos O. Chemical Composition of Bee Pollen. *Applications of Honeybee Plant-Derived Products*, vol. 1, Sharjah, United Arab Emirates: Bentham Books; 2016, p. 67–88.
[13] Campos M, Bogdanov S, Almeida-Muradian L, Szczesna T, Mancebo Y, Frigerio C, et al. Pollen composition and standardisation of analytical methods. *Journal of Apicultural Research and Bee World 2008*;47:156–63. https://doi.org/10.3896/IBRA.1.47.2.12.
[14] Institute of Medicine. *Dietary Reference Intakes for Energy, Carbohydrate, Fiber, Fat, Fatty Acids, Cholesterol, Protein, and Amino Acids.* Washington, DC: The National Academies Press; 2005. https://doi.org/10.17226/10490.
[15] Institute of Medicine. *Dietary Reference Intakes: The Essential Guide to Nutrient Requirements.* Washington, DC: The National Academies Press; 2006. https://doi.org/10.17226/11537.
[16] Institute of Medicine I. *Dietary Reference Intakes for Calcium and Vitamin D.* Washington, DC: The National Academies Press; 2011. https://doi.org/10.17226/13050.
[17] National Academies of Sciences, Engineering, and Medicine. *Dietary Reference Intakes for Sodium and Potassium.* Washington, DC: The National Academies Press; 2019. https://doi.org/10.17226/25353.
[18] Aylanc V, Falcão SI, Ertosun S, Vilas-Boas M. From the hive to the table: Nutrition value, digestibility and bioavailability of the dietary phytochemicals present in the bee pollen and bee bread. *Trends in Food Science & Technology* 2021;109:464–81. https://doi.org/10.1016/j.tifs.2021.01.042.
[19] Crozier A, Jaganath IB, Clifford MN. Dietary phenolics: chemistry, bioavailability and effects on health. *Nat Prod Rep* 2009;26:1001–43. https://doi.org/10.1039/b802662a.

[20] Tsao R. Chemistry and biochemistry of dietary polyphenols. *Nutrients* 2010;2:1231–46. https://doi.org/10.3390/nu2121231.

[21] Martín-Gómez J, García-Martínez T, Varo MÁ, Mérida J, Serratosa MP. Phenolic compounds, antioxidant activity and color in the fermentation of mixed blueberry and grape juice with different yeasts. *LWT* 2021;146: 111661. https://doi.org/10.1016/j.lwt.2021.111661.

[22] Srećković N, Katanić Stanković JS, Matić S, Mihailović NR, Imbimbo P, Monti DM, et al. *Lythrum salicaria* L. (Lythraceae) as a promising source of phenolic compounds in the modulation of oxidative stress: Comparison between aerial parts and root extracts. *Industrial Crops and Products* 2020;155:112781. https://doi.org/10.1016/j.indcrop.2020.112781.

[23] Chang H-T, Jan C-R, Liang W-Z. Protective effects of a phenolic glycoside compound curculigoside on H2O2-induced oxidative stress and cytotoxicity in normal human breast epithelial cells. *Journal of Functional Foods* 2018;41:171–82. https://doi.org/10.1016/j.jff.2017.12.009.

[24] Presti G, Guarrasi V, Gulotta E, Provenzano F, Provenzano A, Giuliano S, et al. Bioactive compounds from extra virgin olive oils: Correlation between phenolic content and oxidative stress cell protection. *Biophysical Chemistry* 2017;230:109–16. https://doi.org/10.1016/j.bpc.2017.09.002.

[25] Chang T-L, Lin S-W, Wu S, Hong C-M. Regulation of ubiquitin and 26S proteasome mediated by phenolic compounds during oxidative stress. *The Journal of Nutritional Biochemistry* 2013;24:1970–81. https://doi.org/10.1016/j.jnutbio.2013.07.001.

[26] Mărgăoan R, Stranţ M, Varadi A, Topal E, Yücel B, Cornea-Cipcigan M, et al. *Bee Collected Pollen and Bee Bread: Bioactive Constituents and Health Benefits. Antioxidants* (Basel) 2019;8:568. https://doi.org/10.3390/antiox8120568.

[27] Razali FN, Sinniah SK, Hussin H, Zainal Abidin N, Shuib AS. Tumor suppression effect of Solanum nigrum polysaccharide fraction on Breast cancer via immunomodulation. *International Journal of Biological Macromolecules* 2016;92:185–93. https://doi.org/10.1016/j.ijbiomac.2016.06.079.

[28] Kroyer G, Hegedus N. Evaluation of bioactive properties of pollen extracts as functional dietary food supplement. *Innovative Food Science & Emerging Technologies* 2001;2:171–4. https://doi.org/10.1016/S1466-8564(01)00039-X.

[29] Almaraz-Abarca N, Campos M da G, Ávila-Reyes JA, Naranjo-Jiménez N, Herrera-Corral J, González-Valdez LS. Variability of antioxidant activity among honeybee-collected pollen of different botanical origin. *Interciencia* 2004;29:574–8.

[30] Anjos O, Fernandes R, Cardoso SM, Delgado T, Farinha N, Paula V, et al. Bee pollen as a natural antioxidant source to prevent lipid oxidation in black pudding. *LWT* 2019;111:869–75. https://doi.org/10.1016/j.lwt.2019.05.105.

[31] Negri G, Barreto LMRC, Sper FL, Carvalho C, Campos MGR. Phytochemical analysis and botanical origin of Apis mellifera bee pollen from the municipality of Canavieiras, Bahia State, Brazil. *Brazilian Journal of Food Technology* 2018;21:e57566.

[32] Suriyatem R, Auras RA, Intipunya P, Rachtanapu P. Predictive mathematical modeling for EC50 calculation of antioxidant activity and antibacterial ability of Thai bee products. *Journal of Applied Pharmaceutical Science* 2017;7:122–33. https://doi.org/10.7324/JAPS.2017.70917.

[33] Almaraz-Abarca N, da Graça Campos M, Ávila-Reyes JA, Naranjo-Jiménez N, Herrera Corral J, González-Valdez LS. Antioxidant activity of polyphenolic extract of monofloral honeybee-collected pollen from mesquite (*Prosopis juliflora*, Leguminosae). *Journal of Food Composition and Analysis* 2007;20:119–24. https://doi.org/10.1016/j.jfca.2006.08.001.

[34] Mărghitaş LA, Stanciu OG, Dezmirean DS, Bobiş O, Popescu O, Bogdanov S, et al. In vitro antioxidant capacity of honeybee-collected pollen of selected floral origin harvested from Romania. *Food Chemistry* 2009;115:878–83. https://doi.org/10.1016/j.foodchem.2009.01.014.

[35] Gonçalves PJS, Estevinho LM, Pereira AP, Sousa JMC, Anjos O. Computational intelligence applied to discriminate bee pollen quality and botanical origin. *Food Chemistry* 2018;267:36–42. https://doi.org/10.1016/j.foodchem.2017.06.014.

[36] Aylanc V, Tomás A, Russo-Almeida P, Falcão SI, Vilas-Boas M. Assessment of Bioactive Compounds under Simulated Gastrointestinal Digestion of Bee Pollen and Bee Bread: Bioaccessibility and Antioxidant Activity. *Antioxidants* 2021;10:651. https://doi.org/10.3390/antiox10050651.

[37] Maruyama H, Sakamoto T, Araki Y, Hara H. Anti-inflammatory effect of bee pollen ethanol extract from Cistus sp. of Spanish on Origin on carrageenan-induced rat hind paw edema. *BMC Complementary and Alternative Medicine* 2010;10:30. https://doi.org/10.1186/1472-6882-10-30.

[38] Pascoal A, Rodrigues S, Teixeira A, Feás X, Estevinho LM. Biological activities of commercial bee pollens: Antimicrobial, antimutagenic, antioxidant and anti-inflammatory. *Food and Chemical Toxicology* 2014;63:233–9. https://doi.org/10.1016/j.fct.2013.11.010.

[39] Feás X, Vázquez-Tato MP, Estevinho L, Seijas JA, Iglesias A. Organic Bee Pollen: Botanical Origin, Nutritional Value, Bioactive Compounds, Antioxidant Activity and Microbiological Quality. *Molecules* 2012;17: 8359–77. https://doi.org/10.3390/molecules17078359.

[40] Ozcan M. Inhibition of Aspergillus parasiticus NRRL 2999 by pollen and propolis extracts. *J Med Food* 2004;7:114–6. https://doi.org/10.1089/109662004322984806.

[41] Morais M, Moreira L, Feás X, Estevinho LM. Honeybee-collected pollen from five Portuguese Natural Parks: palynological origin, phenolic content, antioxidant properties and antimicrobial activity. **Food Chem Toxicol** 2011;49:1096–101. https://doi.org/10.1016/j.fct.2011.01.020.

[42] Carpes ST, Cabral ISR, Rosalen PL, Alencar SM, Masson ML. Caracterização do potencial antimicrobiano dos extratos de pólen apícola da Região Sul do Brasil. Alimentos e Nutrição Araraquara 2009;20:271–7.

[43] Yildiz O, Can Z, Saral Ö, Yuluğ, E, Öztürk F, Aliyazıcıoğlu R, et al. Hepatoprotective Potential of Chestnut Bee Pollen on Carbon Tetrachloride-Induced Hepatic Damages in Rats. *Evidence-Based Complementary and Alternative Medicine* 2013;2013:e461478. https://doi.org/10.1155/2013/461478.

[44] Šarić A, Balog T, Sobočanec S, Kušić B, Šverko V, Rusak G, et al. Antioxidant effects of flavonoid from Croatian *Cystus incanus* L. rich bee pollen. *Food and Chemical Toxicology* 2009;47:547–54. https://doi.org/10.1016/j.fct.2008.12.007.

[45] Bevzo VV, Grygor'eva NP. [Effect of bee pollen extract on glutathione system activity in mice liver under X-ray irradiation]. *Ukr Biokhim Zh* 1997;69:115–7.

[46] Eraslan G, Kanbur M, Silici S. Effect of carbaryl on some biochemical changes in rats: The ameliorative effect of bee pollen. *Food and Chemical Toxicology* 2009;47:86–91. https://doi.org/10.1016/j.fct.2008.10.013.

[47] Yang Y, Liu M, Li H, Yang Y, Su N, Wu Y, et al. Proteomics analysis of the protective effect of canola (*Brassica campestris* L.) bee pollen flavonoids on the tert-butyl hydroperoxide-induced EA.hy926 cell injury model. *Journal of Functional Foods* 2020;75:104223. https://doi.org/10.1016/j.jff.2020.104223.

[48] Laaroussi H, Bakour M, Ousaaid D, Aboulghazi A, Ferreira-Santos P, Genisheva Z, et al. Effect of antioxidant-rich propolis and bee pollen extracts against D-glucose induced type 2 diabetes in rats. *Food Research International* 2020;138:109802. https://doi.org/10.1016/j.foodres.2020.109802.

[49] Medeiros KCP, Figueiredo CAV, Figueredo TB, Freire KRL, Santos FAR, Alcantara-Neves NM, et al. Anti-allergic effect of bee pollen phenolic extract and myricetin in ovalbumin-sensitized mice. *Journal of Ethnopharmacology* 2008;119:41–6. https://doi.org/10.1016/j.jep.2008.05.036.
[50] Zhao Y, Yan Y, Zhou W, Chen D, Huang K, Yu S, et al. Effects of polysaccharides from bee collected pollen of Chinese wolfberry on immune response and gut microbiota composition in cyclophosphamide-treated mice. *Journal of Functional Foods* 2020;72:104057. https://doi.org/10.1016/j.jff.2020.104057.
[51] Arung ET, Ramadhan R, Khairunnisa B, Amen Y, Matsumoto M, Nagata M, et al. Cytotoxicity effect of honey, bee pollen, and propolis from seven stingless bees in some cancer cell lines. *Saudi Journal of Biological Sciences* 2021. https://doi.org/10.1016/j.sjbs.2021.08.017.
[52] Pinto B, Caciagli F, Riccio E, Reali D, Sarić A, Balog T, et al. Antiestrogenic and antigenotoxic activity of bee pollen from Cystus incanus and Salix alba as evaluated by the yeast estrogen screen and the micronucleus assay in human lymphocytes. *Eur J Med Chem* 2010;45:4122–8. https://doi.org/10.1016/j.ejmech.2010.06.001.
[53] Wan Omar WA, Azhar NA, Harif Fadzilah N, Nik Mohamed Kamal NNS. Bee pollen extract of Malaysian stingless bee enhances the effect of cisplatin on breast cancer cell lines. *Asian Pacific Journal of Tropical Biomedicine* 2016;6:265–9. https://doi.org/10.1016/j.apjtb.2015.12.011.
[54] Tuoheti T, Rasheed HA, Meng L, Dong M sheng. High hydrostatic pressure enhances the anti-proliferative properties of lotus bee pollen on the human prostate cancer PC-3 cells via increased metabolites. *Journal of Ethnopharmacology* 2020;261:113057. https://doi.org/10.1016/j.jep.2020.113057.
[55] Végh R, Csóka M, Sörös C, Sipos L. Food safety hazards of bee pollen – A review. *Trends in Food Science & Technology* 2021;114:490–509. https://doi.org/10.1016/j.tifs.2021.06.016.
[56] Ambrus A, Szenczi-Cseh J, Griff T, Kerekes K, Miklós G, Vásárhelyi A, et al. Food safety assessment of the mycotoxin and pesticide residue contamination of our foods, part 2. Mycotoxins. *Elelmiszervizsgalati Kozlemenyek* 2020;66:2922–49.
[57] Bostan I, Onofrei M, Gavriluţă (Vatamanu) AF, Toderaşcu C, Lazăr CM. An Integrated Approach to Current Trends in Organic Food in the EU. *Foods* 2019;8:144. https://doi.org/10.3390/foods8050144.
[58] Costa CA, Guiné RPF, Costa DVTA, Correia HE, Nave A. Chapter 3 - Pest Control in Organic Farming. In: Chandran S, Unni MR, Thomas S, editors.

Organic Farming, Woodhead Publishing; 2019, p. 41–90. https://doi.org/10.1016/B978-0-12-813272-2.00003-3.

[59] Gierer F, Vaughan S, Slater M, Thompson H, Elmore J, Girling R. A review of the factors that influence pesticide residues in pollen and nectar: Future research requirements for optimising the estimation of pollinator exposure. *Environmental Pollution* 2019;249:236–47. https://doi.org/10.1016/j.envpol.2019.03.025.

[60] Wen X, Ma C, Sun M, Wang Y, Xue X, Chen J, et al. Pesticide residues in the pollen and nectar of oilseed rape (*Brassica napus* L.) and their potential risks to honey bees. Science of The Total Environment 2021;786:147443. https://doi.org/10.1016/j.scitotenv.2021.147443.

[61] Böhme F, Bischoff G, Zebitz CPW, Rosenkranz P, Wallner K. Pesticide residue survey of pollen loads collected by honeybees (Apis mellifera) in daily intervals at three agricultural sites in South Germany. *PLOS ONE* 2018;13:e0199995. https://doi.org/10.1371/journal.pone.0199995.

[62] Nai Y-S, Chen T-Y, Chen Y-C, Chen C-T, Chen B-Y, Chen Y-W. Revealing Pesticide Residues Under High Pesticide Stress in Taiwan's Agricultural Environment Probed by Fresh Honey Bee (Hymenoptera: Apidae) *Pollen. J Econ Entomol* 2017;110:1947–58. https://doi.org/10.1093/jee/tox195.

[63] Tong Z, Duan J, Wu Y, Liu Q, He Q, Shi Y, et al. A survey of multiple pesticide residues in pollen and beebread collected in China. *Sci Total Environ* 2018;640–641:1578–86. https://doi.org/10.1016/j.scitotenv.2018.04.424.

[64] Mullin CA, Frazier M, Frazier JL, Ashcraft S, Simonds R, vanEngelsdorp D, et al. High Levels of Miticides and Agrochemicals in North American Apiaries: Implications for Honey Bee Health. *PLOS ONE* 2010;5:e9754. https://doi.org/10.1371/journal.pone.0009754.

[65] Kostić AŽ, Pešić MB, Mosić MD, Dojčinović BP, Natić MM, Trifković JĐ. Mineral content of bee pollen from Serbia / Sadržaj minerala u uzorcima pčelinjega peluda iz Srbije. *Archives of Industrial Hygiene and Toxicology* 2015;66:251–8. https://doi.org/10.1515/aiht-2015-66-2630.

[66] Morgano MA, Teixeira Martins MC, Rabonato LC, Milani RF, Yotsuyanagi K, Rodriguez-Amaya DB. Inorganic Contaminants in Bee Pollen from Southeastern Brazil. *J Agric Food Chem* 2010;58:6876–83. https://doi.org/10.1021/jf100433p.

[67] Álvarez-Ayuso E, Abad-Valle P. Trace element levels in an area impacted by old mining operations and their relationship with beehive products. *Sci Total Environ* 2017;599–600:671–8. https://doi.org/10.1016/j.scitotenv.2017.05.030.

[68] Lambert O, Piroux M, Puyo S, Thorin C, Larhantec M, Delbac F, et al. Bees, honey and pollen as sentinels for lead environmental contamination. *Environ Pollut* 2012;170:254–9. https://doi.org/10.1016/j.envpol.2012.07.012.

[69] Al Naggar Y, Naiem E, Mona M, Seif A. Honey bees and their products as a bioindicator of environmental pollution with heavy metals. MELLIFERA 2013;13:10–20.

[70] Bargańska Ż, Ślebioda M, Namieśnik J. Honey bees and their products: Bioindicators of environmental contamination. *Critical Reviews in Environmental Science and Technology* 2016;46:235–48. https://doi.org/10.1080/10643389.2015.1078220.

[71] Roman A. Concentration of Chosen Trace Elements of Toxic Properties in Bee Pollen Loads. *Pol J Environ Stud* 2009;18:265–72.

[72] Alarcon D, Cáceres E, Aguilera A. Mycotoxins in foods that cause damage to humans. *Journal of Applied Biotechnology & Bioengineering* 2019;6:259–63. https://doi.org/10.15406/jabb.2019.06.00202.

[73] Kostić AŽ, Milinčić DD, Petrović TS, Krnjaja VS, Stanojević SP, Barać MB, et al. Mycotoxins and Mycotoxin Producing Fungi in Pollen: Review. *Toxins* (Basel) 2019;11:64. https://doi.org/10.3390/toxins11020064.

[74] Boppré M, Colegate SM, Edgar JA, Fischer OW. Hepatotoxic Pyrrolizidine Alkaloids in Pollen and Drying-Related Implications for Commercial Processing of Bee Pollen. *J Agric Food Chem* 2008;56:5662–72. https://doi.org/10.1021/jf800568u.

[75] Mädge I, Gehling M, Schöne C, Winterhalter P, These A. Pyrrolizidine alkaloid profiling of four Boraginaceae species from Northern Germany and implications for the analytical scope proposed for monitoring of maximum levels. *Food Addit Contam Part A Chem Anal Control Expo Risk Assess* 2020;37:1339–58. https://doi.org/10.1080/19440049.2020.1757166.

[76] Moreira R, Pereira DM, Valentão P, Andrade PB. Pyrrolizidine Alkaloids: Chemistry, Pharmacology, Toxicology and Food Safety. *Int J Mol Sci* 2018;19:1668. https://doi.org/10.3390/ijms19061668.

[77] Nanotte-Varly C. Why not evaluate the allergenic potential of bee pollen with a skin testing method? *Ann Allergy* 2016;3:114–22.

[78] Abrol DP. Genetically Modified Plants and Bees. In: Abrol DP, editor. *Pollination Biology: Biodiversity Conservation and Agricultural Production*, Dordrecht: Springer Netherlands; 2012, p. 669–707. https://doi.org/10.1007/978-94-007-1942-2_20.

[79] Johnston P, Huxdorff C, Simon G, Santillo D. *The bees' burden: An analysis of pesticide residues in comb pollen (beebread) and trapped pollen*

from honey bees (apis mellifera) in 12 european countries. Exeter, United Kingdom: Greenpeace Research Laboratories; 2014.
[80] Satta A, Verdinelli M, Ruiu L, Buffa F, Salis S, Sassu A, et al. Combination of beehive matrices analysis and ant biodiversity to study heavy metal pollution impact in a post-mining area (Sardinia, Italy). *Environ Sci Pollut Res* 2012;19:3977–88. https://doi.org/10.1007/s11356-012-0921-1.
[81] Garcia-Villanova RJ, Cordón C, González Paramás AM, Aparicio P, Garcia Rosales ME. Simultaneous Immunoaffinity Column Cleanup and HPLC Analysis of Aflatoxins and Ochratoxin A in Spanish Bee Pollen. *J Agric Food Chem* 2004;52:7235–9. https://doi.org/10.1021/jf048882z.
[82] Roman A, Popiela-Pleban E, Migdał P, Kruszyński W. As, Cr, Cd, and Pb in Bee Products from a Polish Industrialized Region. *Open Chemistry* 2016;14:33–6. https://doi.org/10.1515/chem-2016-0007.
[83] Lambert O, Piroux M, Puyo S, Thorin C, L'Hostis M, Wiest L, et al. Widespread Occurrence of Chemical Residues in Beehive Matrices from Apiaries Located in Different Landscapes of Western France. *PLOS ONE* 2013;8:e67007. https://doi.org/10.1371/journal.pone.0067007.
[84] Dübecke A, Beckh G, Lüllmann C. Pyrrolizidine alkaloids in honey and bee pollen. *Food Additives & Contaminants: Part A* 2011;28:348–58. https://doi.org/10.1080/19440049.2010.541594.
[85] European Commission. *EU Pesticides Database 2021.* https://ec.europa.eu/food/plants/pesticides/eu-pesticides-database_en (accessed October 14, 2021).

Reviewed by:

Professor Kemilla Rebelo, Institute of Health and Biotechnology, Federal University of Amazonas, Coari-AM.

Professor Wilma Spinosa, University of Londrina, Londrina-PR, Brazil.

Chapter 7

The Mead as an Alternative Medicinal Treatment

Lucia Perales Serna,
Adriana Carolina Flores-Gallegos, Anilu Rubio Rios,
Juan Alberto Ascacio-Valdés, Aidé Sáenz Galindo
and Raúl Rodríguez-Herrera*
Facultad de Ciencias Químicas, Universidad Autónoma de Coahuila,
Boulevard Venustiano Carranza and José Cárdenas s/n,
Republica Oriente, Saltillo, Coahuila, México

Abstract

Nowadays, the development of foods that promote health and nutrition is one of the priorities in the food industry. Although there is a great variety of products and beverages with a pleasant flavor and nutritional value that have been undervalued, among them is aguamiel (sap from the maguey *Agave* spp), which is a beverage that provides different nutrients and has a cultural tradition in some communities in Mexico. Aguamiel can be considered a unique and special beverage, especially because of its artisanal production process. However, this traditional beverage has been losing its cultural value. On the other hand, aguamiel is the sap of the maguey (*Agave* spp), which is considered a natural resource of great diversity of the Agavaceae family. The human population currently consumes nutritional products and beverages, some are of good quality and provide nutrients, but others do not, since they are highly processed

* Corresponding Author's E-mail: raul.rodriguez@uadec.edu.mx.

In: Food Processing
Editors: Myriam Huijs
ISBN: 978-1-68507-570-5

foods and this causes the quality and nutrition of the product to be low. This reflects the importance of preserving the customs of a traditionally 100% Mexican beverage that is also healthy. In northern Mexico, although there is a tradition of consuming this drink, it could be said that its consumption is minimal, although studies have shown that this drink is healthy, with a delicious flavor, contains vitamin A, B, C and iron, proteins, carbohydrates and is an excellent antioxidant. In addition, this drink has therapeutic medicinal properties, and helps to lower cholesterol. That is why aguamiel is one of the products that has not been fully exploited, since the main use that has been given to it is for the production of pulque, a traditionally Mexican alcoholic beverage that results from the fermentation of aguamiel.

Keywords: Agave salmiana, mead, probiotic, cholesterol

Introduction

In Mexico there are natural resources that have been used to prepare different beverages and foods of pleasant taste and nutritional value, for which this country is recognized for its great gastronomic culture. An example of these resources is mead, which has a high nutritional content because it contains vitamins, amino acids and fatty acids that make it a functional food (Rodrigo Guzmán-Pedraza). In addition, it is a refreshing drink with a sweet flavor, amber color and herbaceous odor, and is obtained from the maguey pulquero. It was a very common product at the beginning of the 20th century due to its great nutritional properties, but over time it gradually lost its recognition. However, the tradition of consuming it is still alive in some places such as Zacatecas, San Luis Potosí and Mexico City.

Aguamiel, also known as syrup or syrup, is obtained from the maguey, a plant that belongs to the Agavaceae family and grows in both mountainous areas and Mexican plains (García-Herrera et al., 2010). Maguey is a plant with thick leaves called pencas, which grow in a cold and semi-dry climate (Segura et al., 2006). To obtain the juice, the maguey must be eight to twelve years old. Mead is the sap that is extracted from the pineapple of the maguey pulquero, its shelf life is approximately two hours at room temperature (Espíndola-Sotres et al, 2018).

On the other hand, mead due to its nutritional properties can be used as a food supplement, it is diuretic, contains probiotics, and can also be considered as a low glycemic index drink (Segura et al., 2006). In addition, it is a great prebiotic, helping to combat gastritis and constipation (Escalante et al., 2010). This drink is also a good antioxidant because it helps strengthen the immune system, and has an alkaline pH close to 8 (Espindola Sotres et al., 2018).

Process of Obtaining

The origin of mead dates back to pre-Hispanic times, as maguey was an extremely important plant in Mesoamerican cultures before the arrival of the Spanish. The maguey was used to weave parts of clothing, and the mead was used as a sweetener (López Mancilla et al, 2007), but its most important function was medicinal and food. This plant is classified as monocarpic, and flowers only once in its life. After this process the plant dies (Muñíz et al., 2013). The leaves that surround the trunk are called pencas, which are distributed in a circular shape and are green or yellowish in color, thick and fleshy, and at the end the leaf ends in a sharp point. The type of climate where this plant develops is in arid or semi-arid zones. (Alfaro-Rojas et al., 2007).

Mead is obtained from adult and mature magueys, it has a watery consistency with a sweet taste, that is why it is preferable to be consumed as soon as possible after being extracted, because after 24 to 48 hours the mead ferments (Flores et al., 2008) due to native microorganisms and yeasts such as: Lactobacillus, Leuconostoc and Zymomonas mobillis, this process gives rise to pulque. The traditional process for obtaining mead is by performing a cutting or capado which consists of removing the tender leaves from the center of the mature plant (Guzmán Pedraza et al., 2018), and a cavity is left where a few days later the sap is collected from the plant. If the cut is made before the ripening time, low mead yields can be obtained. The resulting hole should be covered with agave leaves or a stone to avoid exposure to the environment or possible contamination (Escalante et al., 2012).

Finally, the mead is transferred to containers such as a leather bag or wooden box. The process of collecting the mead is known as "tlachiqueo," where an instrument called "acocote" is used to fill the sap by suction and transfer it to the container (Escalante et al., 2012).

Figure 1 clearly shows the traditional process for obtaining mead.

Figure 1. Photographs of the process of obtaining aguamiel in rural communities of Mexico. a) Mature agave plant; b) Cutting of agave; c) Cavity where aguamiel is accumulated; d) Extraction of aguamiel; e) Deposit of aguamiel in containers; f) The agave dies.

Nutritional Composition

Mead is a hydrating beverage due to its high water content, therefore it helps to replenish liquids as well as providing vitamins and is attractive due to its herbaceous odor and characteristic flavor (NMX-V-022, 1972). In addition, mead is made up of water, sugars, proteins, vitamins and mineral salts and has several types of carbohydrates including fructose, sucrose and glucose, (Valadez et al., 2012), as well as some amino acids and fructans such as inulin (Espindola Sotres et al., 2018). Due to its nutritional properties it has been mentioned as a prebiotic product and this further enriches the product. Table 1 shows the physicochemical composition and minerals present in mead.

Mead can be an important source of nutritional elements for human consumption, its chemical composition is illustrated in Table 2.

Table 1. Physicochemical and mineral composition of mead (Flores et al., 2008, Bautista and Arias et al., 2008)

Chemical composition of mead	**In 100gr. of dry base**
Humidity	89.61%
pH	6.29 – 7.72
Solubles solids (°Brix)	11.10 – 12.67
Titratable acidity	0.06 (%)
Total proteins	3.50 (%)
Total fructose	32.63 (%)
Total glucose	28.68 (%)
Fructooligosacharidess	15.51 (%)
Total sucrose	12.90 (%)
Minerals	
Potassium	120.44 (mg.)
Calcium	11.70 (mg.)
Lead	0.015 (mg.)
Zinc	0.18 (mg.)
Iron	0.81 (mg.)
Sodium	0.83 (mg.)
Copper	0.07 (mg.)
Magnesium	0.55 (mg.)
Selenium	0.047 (mg.)

Table 2. Chemical composition of mead (Cravioto et al., 1953; Loyola 1956; Ramírez-Higuera 2009)

Component	**Mg/L**
Density	10.49
Acidity	0.680
Glucose	0.120
Saccharose	94.50
Rubbers	6.000
Albuminoids	8.060
Ashes	4.500

It is worth mentioning that the components of aguamiel vary from one variety of maguey to another or from one region to another, and depending on the time of year, minimal differences can be found in its composition, as shown in Table 3, which shows the composition of aguamiel extracted from three varieties of maguey.

Table 3. Composition of mead from three maguey varieties (Flores et al., 2008)

	Manzo	*Cenizo*	*Amarillo*
Density (g/L)	1.29	1.26	1.23
Ph	6.30	6.40	6.60
Refractive Index	1.35	1.35	1.36
Soluble solids (Brix)	11.44	11.01	12.67
Acidity (%)	1.65	1.41	1.47
Humidity (%)	87.00	87.90	86.00
Protein (g/L)	3.41	3.11	2.49
Ashes (g)	0.53	0.41	0.48
Reducing sugars(g/L)	1.63	1.97	1.06
Glucose (mg/L)	2.31	2.31	2.50
Fructose (mg/L)	4.70	4.92	4.50

Probiotics in Mead

Biotechnological studies using different molecular techniques based on DNA have reported different microorganisms present in mead, most of them recognized as beneficial or probiotics (G. Oliveira Fuster et al., 2007). Probiotics are live microorganisms that, when administered in adequate amounts, provide a health benefit to the host. One of the main functions of the intestinal flora is to protect the organism from pathogenic microorganisms. Having a well-balanced and varied intestinal flora is good for health. The incorporation of these benign microorganisms or probiotics, help to maintain a good intestinal balance preventing different diseases. However, the probiotic properties depend on some specific strains since their use is recommended preferably under medical indication (Olvera, et al., 2007). Most of the probiotics marketed in foods are of the bacterial genera Lacobacillus and Bifidobacterium as well as the yeast genus Sacharomyces (C. Ramirez et al., 2011).

Some benefits of probiotics are:

- They contribute to restore the microbiota, i.e., the microorganisms that inhabit the intestine.
- They produce antimicrobial substances
- Stimulate the immune system
- Reduce lactose intolerance, i.e., promote digestion at intestinal level.
- Increase resistance to infection by pathogenic organisms in the gut
- Improve metabolism, combating obesity and diabetes.
- Help stress, avoiding anxiety and depression.
- Prevents colds
- Combat the side effects of antibiotics, protecting the natural flora
- Relieves irritable bowel, with less pain, less bloating and less flatulence

Lactic Acid Bacteria

Lactic acid bacteria are a group of microorganisms represented by several genera, with physiological, morphological and metabolomic characteristics in common. They produce lactic acid as a product of carbohydrate fermentation, most of them grow at a pH of 4 to 4.5 (Escalante et al., 2012). Lactic acid bacteria are complemented by bacteria present in the intestinal flora which contribute to the proper functioning of the intestine. Lactic acid bacteria have a high biotechnological potential due to their fermentative process (Ramirez et al., 2011) in foods that are practically destined for human consumption such as vegetables, dairy products, meat, bakery products and alcoholic beverages. Lactic acid bacteria contribute to the development of the organoleptic and rheological characteristics of foods.

Microbial Diversity of Mead

Microorganisms are of great importance in mead. In the fermentation of this beverage, microorganisms such as yeasts, lactic bacteria, ethanol-producing bacteria and bacteria that produce expolysaccharides such as β-glucans and dextrans, which increase the viscosity of the beverage, are found (Cervantes et al., 2007). There are microorganisms that are in charge of naturally transforming sugars into mead, such as: lactic acid producing bacteria: 1) Leuconostoc and Lactobacillus, which increase the acidity of the beverage; 2) non-Saccharomyces and Saccharomyces yeasts and bacteria (Zymomonas mobilis ssp. Mobilis) that transform sugars into ethanol and other secondary metabolites; 3) dextran-producing bacteria (Leuconostoc spp.) that give viscosity to pulque; 4) acetic bacteria (Acetobacter spp.) that together with lactic acid bacteria acidify the beverage (Lape-Olivares et al., 2008).

Medicinal Properties of Mead

Mead can be used to obtain polysaccharides and some high fructose syrups (Valadez et al., 2012). Fructose syrup can be used as a food additive and cause health benefits by increasing iron absorption in children, decreasing the glycemic index and helping to prevent colon cancer (García, et al., 2009).

The agaves have in their composition many fructans that are inulin type of low caloric value and that is why they can be used for the development of functional foods (González et al., 2012). Mead has nutritional and physicochemical properties since positive effects have been attributed to it, such as strengthening defense mechanisms (Velázquez et al., 2014), improving lipid metabolism, cholesterol, glucose, controlling obesity, preventing osteoporosis (Urias et al., 2007).

Importance in Nutrition

Mead can be considered as a food, since among its components can be found vitamins, minerals, protein, calcium and fiber. (Flores, et al., 2008). This drink has a sweet taste due to its fructose content, and is rich in carbohydrates such as inulin and sucrose. In pre-Hispanic times, maguey nectar was the most widely used sweet. The Aztecs used most of the mead to produce maguey nectar, since they also used it as an energetic and healing remedy for several diseases, but with the arrival of the Spaniards, agave honey was replaced by cane sugar.

Mead is an antioxidant, contains nutritional compounds such as vitamin C, iron, calcium, sodium and potassium, the carbohydrates (Dominguez, et al.,

2008) that this drink has are very delicious, it is rich in fiber, has therapeutic medicinal properties that help control bad cholesterol.

Regular consumption of mead cleanses and improves the functioning of the digestive system. Among the uses of mead is to strengthen the body's defenses and is ideal for diabetics, since the sugars it contains help lower excess sugar in the blood.

Medicinal Alternative

Mead really has nutritional attributes since mead has properties that inhibit the growth of pathogenic bacteria and stimulates the growth of intestinal flora. It can be taken on an empty stomach, reduces insomnia, is a good nutritional supplement and fights anemia.

The niacin contained in mead allows it to drain and clean veins and arteries. It helps intestinal motility and is recommended for people with constipation. It increases the absorption of calcium and magnesium helping osteoporosis (Mataix et al., 2009).

In current studies, inulin has been found (Silos et al., 2007). Mead contains water, fructose, minerals such as iron and zinc in quantities of 21,500 and 14,100 mg/l respectively (Loyola, 1956: Cravioto and Mssieu 1951).

Mead contains fructoligosaccharides which help to improve the digestive system and the elimination of toxins and fats that damage the human body (Matarese et al., 2004). Currently, inulin and fructoligosaccharides are considered as functional foods, which is why it is important to use mead as a food because it provides health benefits to those who consume it.

References

Campos, I., Escalante, A., & Giles-Gómez, M. 2010. *Aislamiento e identificación de bacterias lácticas del pulque con capacidad probiótica.* Universidad Nacional Autónoma de México. Ciudad. [Isolation and identification of lactic acid bacteria from pulque with probiotic capacity.]

Cravioto, R., y Massieu, G. (1951). *Composición de alimentos mexicanos*, pp. 15-16. [*Mexican food composition*]

Delgado Javier Orlando, *Aplicación de Ultrasonido en la Industria de los Alimentos, Tecnología e Ingeniería*, Universidad Nacional Abierta y a

Distancia-UNAD, Bogota, 2011. [*Ultrasound Application in Food Industry, Technology and Engineering*]

Elizabeth Naranjo Gómez por Editor Alimentos, 2015. *Revista, bebidas funcionales "Una necesidad saludable."* [*Magazine, functional drinks "A healthy necessity."*]

Espíndola-Sotres, V., Trejo-Márquez M. A., Lira-Vargas A., Pascual-Bustamante, S. (2018). Caracterización de aguamiel y jarabe de agave originario del Estado de México, Hidalgo y Tlaxcala. *Investigación y Desarrollo en Ciencia y Tecnología de Alimentos* 3: 522-528. [Characterization of mead and agave syrup from the State of Mexico, Hidalgo and Tlaxcala. *Research and Development in Food Science and Technology*]

Flores Morales, A., Mora Escobedo, R., & Romero Aguilar, L. (s.f.). *Evaluación fisicoquímica del aguamiel de tres variedades de maguey pulquero (Agave spp).* Tlaxcala, México. [*Physicochemical evaluation of the mead of three varieties of maguey pulquero (Agave spp).*]

Olveira Fuster G. e I. González-Molero Unidad de Nutrición Clínica y Dietética. Servicio de Endocrinología y Nutrición. *Hospital Regional Universitario Carlos Haya*. Málaga. Instituto de Salud Carlos III. Red RD06/001510008. España. [*Carlos Haya Regional University Hospital.*]

García M., D. R. (2014). Agave Fructans: Their Effect on Mineral Absorption and Bone Mineral Content. *Journal of medicinal food, 17*(11), 1-9.

García, H. y Vázquez, R. (1998). *Cuantificación de Proteínas: una revisión (Bitácora).* Instituto de Biotecnología UNAM. Morelos, México.

García, M., V. A. Sáenz, M. A. Rodríguez, F. J. Vicente, E. Botello, H. Jiménez, M. Cárdenas, R. Rico y J. L. Navarrete. 2009. Strategy for Biotechnological Process Desing Applied to the enzymatic hydrolysis of Agave Fructooligosaccharides to obtain fructose-rich syrups. *J. Agr. Food Chem.* 57: 10205-10210.

González, L., M. E. Jaramillo, A. Bernardino y R. Mora. 2012. Influence of the harvest age on fructan content and fructosyltransferase activity in Agave atrovirens karw pine. International *J. Plant Physiol. Biochem.* 4: 110-119.

Granados S., D. (1993.). *Los agaves en México*. Universidad Autónoma de Chapingo. [*Agaves in Mexico*. Autonomous University of Chapingo.]

Guillermo Salamanca G., Mónica Patricia Osorio T., Leidy Marcela Montoya Grupo de Investigaciones Mellitopalinológicas y Propiedades Fisicoquímicas de Alimentos, Facultad de Ciencias Universidad del Tolima. Tolima, Colombia. *Rev. Chil Nutr.* Vol. 37, No 1 2010. [Leidy Marcela Montoya Group for Mellitopalynological Research and Physicochemical Properties of Food, Faculty of Sciences, University of Tolima. Tolima, Colombia. *Rev. Chil Nutr.*]

Henry Jurado-Gámez, Javier Martínez-Benavides, Jairo Alexánder Morillo-Garcés, Darío Alejandro Romero-Benavides, Adriana Elisabeth Orbes-Villacorte, Laura Nathaly Mesías-Pantoja. *Determinación de la cinética de fermentación en dos medios probióticos, pruebas de desempeño in vitro y efecto de inhibición de Lactobacillus plantarum.* Colombia 2016. [*Determination of fermentation kinetics in two probiotic media, in vitro performance tests and inhibition effect of Lactobacillus plantarum.*]

Ramírez Ramírez José Carmen, Petra Rosas Ulloa, Martha Yanira Velázquez, González, José Armando Ulloa, Francisco Arce Romero, *Bacterias Lácticas: Importancia en alimentos y su efecto en la salud.* Unidad Académica de Medicina. [*Lactic bacteria: Importance in food and its effect on health.*]

Calderón Juan Fernando Vargas Tesis. Ajuste de un modelo cinético para el crecimiento de Lactobacillus acidophilus. Fundación Universidad de América en la facultad de ingenierías. Departamento de ingeniería química Bogotá 2017. Directora Adriana Inés Páes Morales. [*Fit of a kinetic model for the growth of Lactobacillus acidophilus.* Fundación Universidad de América in the engineering faculty.]

Kolida S., G. G. (2007). Prebiotic Capacity of Inulin-Type Fructans. *The journal of nutrition, 07*, 2503-2506.

Lappe-Oliveras P., Moreno-Terrazas R., Arrizón-Gaviño J. Herrera-Suárez T., García-Mendoza A. y Gschaedler-Mathis A. 2008. Yeasts associated with the production of Mexican alcoholic nondistilled and distilled Agave beverages. *FEMS Yeast Res.* 8: 1037-1052. Escuela Ciencias Básicas

López, M. J.; Nichols, N. N.; Dien, B. S.; Moreno, J.; Bothast, R. J. (2004). Isolation of microorganisms for biological detoxification of lignocellulosic hydrolysates. *Applied Microbiology and Biotechnology*, 64(1):125-131.

Loyola, M. (1956). *La industria del pulque*. México: Banco de México. [*The pulque industry*]

Madrigal L., S. E. (2007). La inulina y derivados como ingredientes claves en alimentos funcionales. *Archivos latinoamericanos de la nutrición, 57*(4), 387-396. [Inulin and derivatives as key ingredients in functional foods. *Latin American Archives of Nutrition*]

Maguey-Pulquero (*Agave mapisaga*) throughout the Harvest Period. *Journal of Agricultural and Food Chemistry*, 56, 3682-3687.mbia pp. 135 – 146.

María Fernanda Espinoza Acevedo; Martín-Gil J., Martín-Gil F.J, José-Yacamán M., Carapia-Morales L. and Falcón-Bárcenas T. Microwave-assisted Synthesis of Hydrated Sodium Uranyl Oxonium Silicate. *Polish Journal of Chemistry*. 2005. 79, 1399-1403.

María. 2016, El maguey pulquero, una planta multifuncional y polifacética: los usos desde una visión mestiza e indígena. *Scripta Ethnologica*, 38:65-87.

[The pulquero maguey, a multifunctional and multifaceted plant: the uses from a mestizo and indigenous perspective. *Scripta Ethnologica*]

Mario Cervantes Contreras / Aura Marina Pedroza Rodríguez. *El pulque, características microbiológicas y contenido alcohólico mediante espectroscopia reman*. Nova, julio-diciembre, año 2007/vol. 5, número 008. Universidad Colegio Mayor de Cundinamarca, Bogotá, Colo. [*Pulque, microbiological characteristics and alcohol content by reman spectroscopy*]

Martín, A., Serrano, S., Santos, A., Marquina, D., & Vázquez, C. (2010). Bioluminiscencia bacteriana. *Reduca, 3*(5), 75-86.

Mataix, J. (2009). *Nutrición y alimentación humana*. Barcelona: Océano/Ergon. [*Nutrition and human food*]

Matarese, L. (2004). *Nutrición clínica práctica.* España: Elsevier. [*Practical clinical nutrition*]

Muñiz D, R. R. (2013). Producción Artesanal del Aguamiel: Una Bebida Tradicional Mexicana. *Revista Científica de la Universidad Autónoma de Coahuila, 5*(10), 12-19. [Artisan Production of Mead: A Traditional Mexican Drink. *Scientific Journal of the Autonomous University of Coahuila*]

Muñiz-Márquez, D. B., Rodríguez-Jasso, R. M., Rodríguez-Herrera, R., Contreras Esquivel, J. C., Aguilar-González, C. N. (2013). Producción Artesanal del Aguamiel: Una Bebida Tradicional Mexicana. *Revista Científica de la Universidad Autónoma de Coahuila* 5(10). [Artisan Production of Mead: A Traditional Mexican Drink. *Scientific Journal of the Autonomous University of Coahuila*]

Lappe-Oliveras Patricia, Carmen Wacher-Rodarte. *Identificación polifásica de levaduras y bacterias ácido-lácticas aisladas de aguamiel, pulque y semilla.* Centro de Investigación Científica Superior de Ensenada. 2008. Tesis. Mildred Herrera-Solórzano*, [*Polyphasic identification of yeasts and lactic acid bacteria isolated from mead, pulque and seed.* Ensenada Superior Scientific Research Center]

Ramirez, A. (Noviembrede 2009) Evaluación del efecto probiótico del aguamiel de Maguey (Agave Salmiana) en *Lactobacillus del brueckiisub sp. Bulagricus.* Instituto Politécnico Nacional. *Unidad Profesional Interdisciplinaria de Biotecnología.* México D.F., pp.4-14. [Evaluation of the probiotic effect of Maguey mead (Agave Salmiana) on *Lactobacillus* from brueckiisub sp. Bulagricus. National Polytechnic Institute. Interdisciplinary Professional Unit of Biotechnology]

Segura, J. (2006). El maguey, memoria sobre el cultivo y beneficio de sus productos. *Revista de Geografía Agrícola (julio-diciembre).* ISSN 0186-4394. Consultado el 13 de agosto de 2016. Disponible en: <http://www.redalyc.org/articulo.oa?id=75703709> . [The maguey, memory of the cultivation and benefit of its products. *Journal of Agricultural Geography* (July-December).]

Silos E. H., N. González C., A. Carrillo L., F. Guevara L., M. E. Valverde G. y O. Paredes L. (2005). *Composición química de aguamiel y pencas de Agave salmiana* Gentry. V Congreso del Noroeste, I Nacional, en Ciencias Alimentarias y Biotecnología Centro de las Artes de la Universidad de Sonora Hermosillo. [*Chemical composition of mead and leaves of Agave salmiana* Gentry. V Northwest Congress, I National, in Food Sciences and Biotechnology Center of the Arts of the University of Sonora Hermosillo.]

Silos Espino, H., Tovar Robles, C.L., González Cortés, N., Méndez Gallegos, S., & Rossel Kipping, D. (12-13 de Noviembre de 2010). *Estudio integral del maguey (Agavesalmiana): propagación y valor nutricional.* México: Universidad Autónoma de Nuevo León. [*Comprehensive study of the maguey (Agavesalmiana): propagation and nutritional value*.]

Silos, E. (2007). "Chemical composition and in vitro propagation of Agave salmiana Gen-try," *The Journal of Horticultural Science and Biotechnology*, pp. 355-359.

Tovar, L. R., M. Olivos y M. E. Gutierréz, 2008, Pulque an alcoholic drink from rural México, contains phytase. Its in vitro effects on corn tortilla. P*lant Food Hum Nutr*, 63:189-194.

Urías J., C. P. (2007). Physiological effects of dietary fructans extracted from *Agave tequilana* Gto. and *Dasylirion* spp. *British Journal of nutrition, 13*(7), 2-7.

Vázquez García, Adriana; Aliphat Fernández, Mario Manuel; Estrella Chulim, Néstor Gabriel; Ortiz Torres, Enrique; Ramírez Juárez, Javier; Ramírez, Andrés.

Velázquez J., G. R. (2014). Prebiotic Potential of *Agave angustifolia* Haw Fructans with Different Degrees of Polymerization. *Molecules, 19*, 12660-12675.

Links

http://www.mexicodesconocido.com.mx/mezcales.html
http://www.mezcalpiedra.com
http://www.todomezcal.com/historia.htm
http://www.revistaeluniversal.com.mx

Index

L

M

N

O

P

Q

R

S